Essential NMR

Bernhard Blümich

Essential NMR

For Scientists and Engineers

Second Edition

 Springer

Bernhard Blümich
Institut für Technische und
Makromolekulare Chemie
RWTH Aachen University
Aachen, Germany

ISBN 978-3-030-10703-1   ISBN 978-3-030-10704-8   (eBook)
https://doi.org/10.1007/978-3-030-10704-8

Library of Congress Control Number: 2019931515

1st edition: © Springer-Verlag Berlin Heidelberg 2005
2nd edition: © Springer Nature Switzerland AG 2019, corrected publication 2019

This Springer imprint is published by the registered company Springer Nature Switzerland AG
The registered company address is: Gewerbestrasse 11, 6330 Cham, Switzerland

Preface to the Second, Revised Edition

NMR is unique in many ways. Among the diverse analytical techniques it is one of the most lively ones despite being 70 years of age. Technological progress and serendipitous discoveries are the main drivers. More than a decade has passed since the first edition of this book was completed. During this time the focus on the essentials of NMR has shifted. In particular, the interest in solid-state NMR imaging has faded, while one- and two-dimensional Laplace NMR have been established as viable tools to characterize soft matter and porous media. Hyperpolarization techniques find increasing attention for enhancing the weak thermodynamic nuclear polarization. Lastly, the commercially available NMR instruments have diversified.

Today, compact instruments are on the market that are suitable not only for materials testing by relaxation and diffusion measurements but also for chemical analysis by high-resolution spectroscopy. Although their magnetic fields are as low as those of the early NMR spectrometers, their analytical power is much better, because they provide nearly the same methodology as today's high-field instruments. They can be used on the chemical workbench during synthesis or as sensors for controlling chemical and technological processes. A forerunner of this technology are welllogging tools, which are mobile instruments in frequent use for characterizing the fluids and rock formations of the walls from oil wells. All of these facets of NMR are now treated in the extensively revised 2nd edition of 'Essential NMR'. I am grateful to Gerlind Breuer and Kawarpal Singh for invaluable help with proof-reading an editing.

Aachen, Germany   Bernhard Blümich
September 2018

The original version of the book was revised: Missed out author corrections have been incorporated. The correction to the book is available at https://doi.org/10.1007/978-3-030-10704-8_7

Preface to the First Edition

NMR means Nuclear Magnetic Resonance. It is a phenomenon in physics, which has been exploited for more than 50 years in a manifold of different forms with numerous applications in chemical analysis, medical diagnostics, biomedical research, materials characterization, chemical engineering, and well logging. Although the phenomenon is comparatively simple, the different realizations of NMR in terms of methods to gather molecular information steadily increase following the advances in electronics and data processing.

A scientist or engineer who wants to gain first insight into the basic principles and applications of NMR is faced with the problem of finding a comprehensive and sufficiently short presentation of the essentials of NMR. This is what this book is meant to be. Preferably it is used to accompany a course or to review the material. The figures and the text are arranged in pairs guiding the reader through the different aspects of NMR. Following the introduction, the principles of the NMR phenomenon are covered in Chapter 2. Chapter 3 on spectroscopy addresses the scientist's quest for learning about molecular structure, order, and dynamics. Chapters 4 and 5 deal with imaging and low-field NMR. They are more of interest to the engineer concerned with imaging, transport phenomena, and quality control. It is hoped, that this comprehensive presentation of NMR essentials is a helpful source of information to students and professionals in the applied sciences and in engineering.

Aachen, Germany
May 2004

Bernhard Blümich

Suggested Readings

For selective studies, the following combinations of chapters are recommended:

Topic of interest	Chapters	Readers
NMR spectroscopy	1,2,3	Chemists, physicists, biologists
NMR imaging	1,2,4	Materials scientists, engineers
NMR for quality control	1,2,5	Materials scientists, engineers
NMR physics	1,2,6	Physicists, methodologists

Contents

1 Introduction 1

2 Basic Principles 9

3 Spectroscopy 35

4 Imaging and Transport 73

5 Relaxometry and Laplace NMR 111

6 Hyperpolarization 143

Correction to: Essential NMR C1

Index.. 159

1. Introduction

Uses of NMR
Definition
Equipment
History
Literature

© Springer Nature Switzerland AG 2019
B. Blümich, *Essential NMR*,
https://doi.org/10.1007/978-3-030-10704-8_1

NMR: Nuclear Magnetic Resonance

NMR is a physical resonance phenomenon utilized to investigate molecular properties of matter by irradiating atomic nuclei in magnetic fields with radio waves

Uses of NMR

- *Chemical analysis*: Molecular structures and dynamics
- *Medical diagnostics*: Magnetic resonance imaging of soft matter
- *Chemical engineering*: Quantification of flow and diffusion, reaction monitoring
- *Materials science*: Characterization of soft matter and porous materials
- *Geophysical well logging*: Analysis of pore networks and fluid typing for oil exploration

Magnetic Field Homogeneity and Information

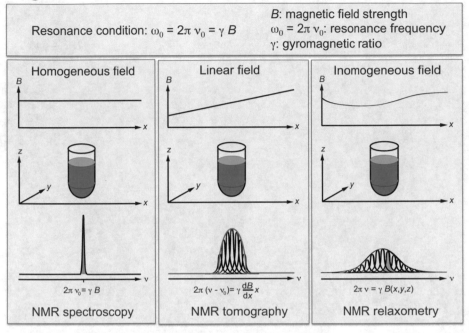

Resonance condition: $\omega_0 = 2\pi\,\nu_0 = \gamma\,B$

B: magnetic field strength
$\omega_0 = 2\pi\,\nu_0$: resonance frequency
γ: gyromagnetic ratio

Homogeneous field

$2\pi\,\nu_0 = \gamma\,B$

NMR spectroscopy

Linear field

$2\pi\,(\nu - \nu_0) = \gamma\,\dfrac{dB}{dx}\,x$

NMR tomography

Inomogeneous field

$2\pi\,\nu = \gamma\,B(x,y,z)$

NMR relaxometry

Types of Information Accessible by NMR

- The most important relationship in NMR states that the *resonance frequency* ω is proportional to the strength B of the *magnetic field*
- Depending on the degree of homogeneity of the applied magnetic field over the extension of the sample, different information can be retrieved by NMR measurements
- If the applied field varies less across the sample than the magnetic fields caused by the electrons surrounding the nuclei, the field is said to be homogeneous. Then NMR spectra can be measured with narrow lines from molecules in solution, which reveal the chemical structure of the molecules. This type of NMR is called *NMR spectroscopy*
- If the applied field varies linearly across the sample, then so does the resonance frequency, and the NMR spectrum is a projection of the distribution of nuclei in the sample along the gradient direction. This is the basis of *magnetic resonance imaging (MRI)* or *NMR tomography* to image objects
- If the applied field varies arbitrarily across the sample, the nuclei in each volume element resonate at a different frequency, and a broad unstructured distribution of resonance frequencies is observed. In such inhomogeneous magnetic fields, NMR relaxation times can be measured, which scale with the physical properties of the samples. They are affected by the molecular mobility such as the elasticity of polymers and the viscosity of liquids. This type of NMR is called *NMR relaxometry*

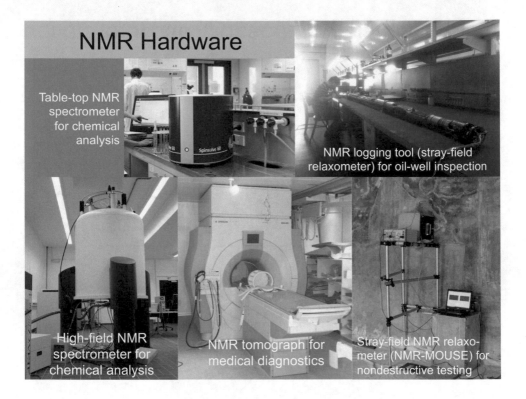

Equipment

- *Spectroscopy* for chemical analysis: *NMR spectrometer* consisting of a magnet with a highly homogeneous field, a radio-frequency transmitter, a receiver, and a computer
- *Tomography / imaging*: *NMR tomograph* consisting of a magnet with a homogeneous field, a modulator for magnetic *gradient fields*, a radio-frequency transmitter, a receiver, and a computer
- *Relaxometry* for *materials testing*: *NMR relaxometer* consisting of a magnet without particular requirements to field homogeneity, preferably a modulator for magnetic *gradient fields*, a radio-frequency transmitter, a receiver, and a computer. NMR relaxometers are typically tabletop instruments with permanent magnets
- *Unilateral NMR*: Stray-field NMR relaxometer like the *NMR-MOUSE®* suitable for *non-destructive testing* of large objects
- *NMR logging*: *NMR relaxometer* incl. stray-field magnet in a tube, shock resistant, and temperature resistant up to 170° C for insertion into a ground hole

Some Nobel Prizes for NMR

Kurt Wüthrich,
1938.
Nobel Prize in
Chemistry 2002

ENC 1995, Boston

Felix Bloch,
1905 - 1983.
Nobel Prize in
Physics 1952

Sir Peter Mans-
field, 1933-2017.
Nobel Prize in
**Medicine or
Physiology 2003**

Richard R. Ernst,
1933.
Nobel Prize in
Chemistry 1991

Edward Mills
Purcell,
1912 - 1997.
Nobel Prize in
Physics 1952

Paul Lauterbur,
1929-2007.
Nobel Prize in
**Medicine or
Physiology 2003**

From Wikimedia: Wüthrich, Bloch: CC BY-SA 3.0;
Mansfield: license GFDL 1.2, Lauterbur: public domain

History of NMR

1945: First successful detection of an NMR signal by Felix Bloch (Stanford)
   and Edward Purcell (Harvard): Nobel Prize in Physics 1952
1949: Discovery of the *NMR echo* by Erwin Hahn
1950: *Indirect spin-spin coupling* by W. G. Proctor and F. C. Yu
1951: *Chemical shift* by J. T. Arnold, S.S. Dharmatti, and M.E. Packard
1966: *Fourier NMR* by Richard Ernst, Nobel Prize in Chemistry 1991
1971: *Two-dimensional NMR* by Jean Jeener, later by Richard Ernst
1972: Proton enhanced high-resolution spectroscopy of dilute spins in solids
   through *cross-polarization* by Alexander Pines, M.G. Gibby, J.S. Waugh
1973: *NMR imaging* by Paul Lauterbur and Peter Mansfield, Nobel Prize in
   Medicine or Physiology 2003
1977: *High-resolution ^{13}C solid-state NMR spectroscopy* with *cross polarization*
   and *magic angle spinning* by J.S. Waugh, E.O. Stejskal, and J. Schaefer
1979: *2D Exchange NMR* by J. Jeener. Application to protein analysis in
   molecular Biology by Kurt Wüthrich, Nobel Prize in Chemistry 2002
1980: *Unilateral NMR* in process control and medicine by Jasper Jackson
1984: *Hyperpolarization* of xenon by William Happer
1995: Commercialization of *well logging NMR*
2005: *Long-lived spin states* by Malcolm Levitt
2009: Contact hyperpolarization with para-hydrogen by Simon Duckett
2010: Tabletop NMR spectroscopy

Recommended Reading: General

- B. Blümich, S. Haber-Pohlmeier, W. Zia, Compact NMR, de Gruyter, Berlin, 2014
- J. Keeler, Understanding NMR Spectroscopy, 2nd ed., Wiley, Chichester, 2010
- M. Levitt, Spin Dynamics, 2nd ed., Wiley, Chichester, 2008
- C.P. Slichter, Principles of Magnetic Resonance, 3rd ed., Springer, Berlin, 1996
- R.R. Ernst, G. Bodenhausen, A. Wokaun, Principles of Nuclear Magnetic Resonance in One and Two Dimensions, Clarendon Press, Oxford, 1987
- E. Fukushima, S.B.W. Roeder, Experimental Pulse NMR: A Nuts and Bolts Approach, Westview, Boulder, 1981
- A. Abragam, The Principles of Nuclear Magnetism, Clarendon Press, Oxford, 1961

Recommended Reading: Relaxometry

- M. Johns, E.J. Friedjonson, S. Vogt, A. Haber, eds., Mobile NMR and MRI, Royal Society of Chemistry, Cambridge, 2016
- F. Casanova, J. Perlo, B. Blümich, eds., Single-Sided NMR, Springer, Berlin, 2011
- J. Kowalewski, L. Mäler, Nuclear Spin Relaxation in Liquids: Theory, Experiments, and Applications, Taylor & Francis, London, 2006
- V.I. Bakhmutov, Practical NMR Relaxation for Chemists, Wiley, Chichester, 2004
- G.R. Coates, L. Xiao, M.G. Prammer, NMR Logging: Principles and Applications, Halliburton Energy Services, Houston, 1999

Recommended Reading: Imaging & Transport

- L. Ciobanu, Microscopic Magnetic Resonance Imaging, Pan Stanford Publishing, Singapore, 2017
- R.W. Brown, Y.-C.N. Chung, E.M. Haacke, M.R. Thompson, R. Venkatesan, Magnetic Resonance Imaging, Physical Principles and Sequence Design, 2nd edition, Wiley, Hoboken, 2014
- E. Hardy, NMR Methods for the Investigation of Structure and Transport, Springer, Berlin, 2012
- P.T. Callaghan, Translational Dynamics & Magnetic Resonance, Oxford University Press, Oxford, 2011
- W.S. Price, NMR Studies of Translational Motion: Principles and Applications, Cambridge University Press, Cambridge, 2009
- B. Blümich, NMR Imaging of Materials, Clarendon Press, Oxford, 2000
- R. Kimmich, NMR Tomography, Diffusometry, Relaxometry, Springer, Berlin, 1997
- M.T. Vlaardingerbroek, J.A. den Boer, Magnetic Resonance Imaging, Springer, Berlin, 1996
- P.T. Callaghan, Principles of Nuclear Magnetic Resonance Microscopy, Clarendon Press, Oxford, 1991

Recommended Reading: Spectroscopy

- V.I. Bakhmutov, NMR Spectroscopy in Liquids and Solids, Taylor & Francis, Boca Raton, 2015
- M. Findeisen, S. Berger, 50 and More Essential NMR Experiments: A Detailed Guide, Wiley-VCH, Weinheim, 2014
- H. Günther, NMR Spectroscopy: Basic Principles, Concepts, and Applications in Chemistry, 3rd edition, Wiley-VCH, Weinheim, 2013
- D. Apperley, R.K. Harris, P. Hodgkinson, Solid-State NMR: Basic Principles and Practice, Momentum, New York, 2012
- R.K. Harris, R.E. Wasylishen, M.J. Duer, eds., NMR Crystallography, Wiley, Chichester, 2009
- M.J. Duer, Solid-State NMR Spectroscopy, Blackwell, Oxford, 2004
- M.J. Duer, ed., Solid-State NMR Spectrocopy, Blackwell, Oxford, 2002
- K. Schmidt-Rohr and H.W. Spiess, Multidimensional Solid-State NMR and Polymers, Academic Press, London, 1994

2. Basic Principles

NMR spectrum
Nuclear magnetism
Rotating coordinate frame
NMR spectrometer
Pulsed NMR
Fourier transformation
Phase correction
Relaxation
Spin echo
Measurement methods
Spatial resolution
Fourier and Laplace NMR

© Springer Nature Switzerland AG 2019
B. Blümich, *Essential NMR*,
https://doi.org/10.1007/978-3-030-10704-8_2

NMR is a Form of Telecommunication in a Magnetic Field

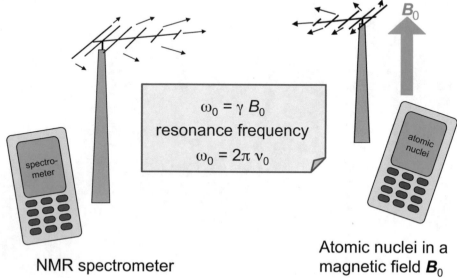

NMR spectrometer

Atomic nuclei in a magnetic field B_0

$$\omega_0 = \gamma\, B_0$$
resonance frequency
$$\omega_0 = 2\pi\, \nu_0$$

Properties of Atomic Nuclei

- Some magnetic isotopes important to NMR are listed below along with their resonance frequencies ν_0 at 1 T field strength, natural abundance, and sensitivity relative to ^{13}C
- Chemical shift range and reference compound are relevant for chemical analysis by NMR spectroscopy
- ^1H is the most sensitive stable nucleus for NMR and the most abundant nucleus in the universe

Nuclear isotope	Natural abundance [%]	Sensitivity rel. to ^{13}C	Spin	ν_0 at 1.0 T [MHz]	Chemical shift range [ppm]	Reference compound
^1H	99.99	5878	½	42.58	12 to −1	$SiMe_4$
^2H	0.01	0.00652	1	6.54	12 to −1	$SiMe_4$
^{13}C	1.07	1.0	½	10.71	240 to −10	$SiMe_4$
^{15}N	0.36	0.0223	½	30.42	1200 to −500	$MeNO_2$
^{19}F	100.00	4890	½	40.06	100 to −300	$CFCl_3$
^{29}Si	4.69	2.16	½	8.46	100 to −400	$SiMe_4$
^{31}P	100.00	391	½	17.24	230 to −200	H_3PO_4

Electrons in Motion

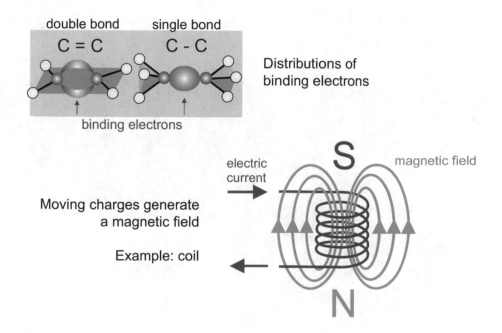

double bond   single bond

C = C   C - C

Distributions of
binding electrons

binding electrons

Moving charges generate
a magnetic field

Example: coil

electric
current

S

magnetic field

N

Magnetic Shielding

- The NMR frequency is determined by the magnetic field at the site of the nucleus
- Atomic nuclei are surrounded by electrons
- In molecules, the electrons of the chemical bond are shared by different nuclei
- Electrons of atoms and molecules move in orbitals, which are studied in quantum mechanics
- The orbitals of the binding electrons are characteristic of the *chemical structure* of the molecule
- Electrons carry an electric charge
- Electric charges in motion generate a magnetic field
- The internal magnetic field generated by the electrons moving in the external magnetic field B_0 is usually opposed to B_0. It shields the nucleus from B_0

Frequency Distributions of Radio Waves

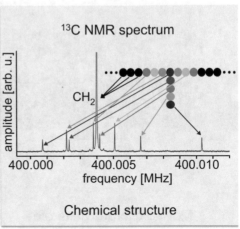

^{13}C NMR spectrum

CH$_2$

Chemical structure

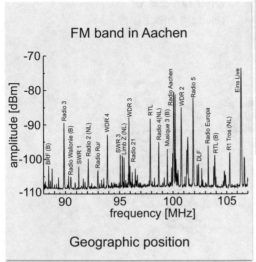

FM band in Aachen

Geographic position

Spectra are fingerprints

Chemical Shift

- The magnetic field generated by the electrons shifts the resonance frequency,
$$\omega_0 = 2\pi\, \nu_0 = \gamma\, (1 - \sigma)\, B_0$$
- The quantity σ is the *magnetic shielding* for a given chemical group
- The quantity $\delta = (\nu_0 - \nu_{ref})\, /\, \nu_{ref}$ is the *chemical shift* of a chemical group. It is independent of the strength B_0 of the magnetic field
- The quantity ν_{ref} is the reference frequency, for example, the resonance frequency of tetramethyl silane (TMS: SiMe$_4$) for ^1H and ^{13}C NMR
- The chemical shift can be calculated from tabulated chemical shift increments as well as *ab initio* from quantum mechanics
- Magnetically inequivalent chemical groups possess different chemical shifts
- In liquids narrow resonance signals are observed with typical widths of 0.1 Hz
- The distribution of resonance frequencies forms the *NMR spectrum*
- The NMR spectrum is an easy-to-read fingerprint of the molecular structure in a way similar to the distribution of FM radio signals at a given location, which provides a fingerprint of the geographic position
- NMR spectra of molecules in solution are measured routinely for chemical identification and structural analysis

Spin and Precession

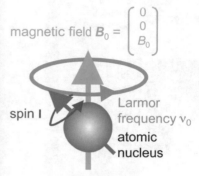

magnetic field $B_0 = \begin{pmatrix} 0 \\ 0 \\ B_0 \end{pmatrix}$

spin I

Larmor frequency ν_0

atomic nucleus

Arnold Sommerfeld, 1868 – 1951. Heisenberg's teacher: Theory of the spinning top (Wikimedia, public domain)

Paul Adrien Maurice Dirac, 1902 – 1984. Nobel prize in Physics 1933: Theory of the spin (Wikimedia, public domain)

Otto Stern 1888 – 1963. Nobel prize in Physics 1943: Experimental proof of the spin (Wikimedia, public domain)

Nuclear Magnetism

- In an NMR sample of material there are of the order of 1 mole or 6×10^{23} atomic nuclei
- 1 mole is 18 g of water. It would take a human 20,000 times the age of the universe to count to one mole
- Some atomic nuclei appear to spin and exhibit an angular momentum
- Examples: ^1H, ^2H, ^{13}C, ^{15}N, ^{19}F, ^{29}Sl, ^{31}P
- Because atomic nuclei are small elementary particles, the laws of classical physics do not apply. Instead the laws of *quantum mechanics* do
- In the laws of physics involving elementary particles *Planck`s constant* $h = 2\pi\, \hbar$ appears
- According to quantum mechanics an elementary particle with an *angular momentum* $\hbar\, I$ or *spin I* also possesses a magnetic dipole moment $\mu = \gamma\, \hbar\, I$
- A classical object with angular momentum is the spinning *bicycle wheel*
- A wheel spinning in a gravitational field formally follows the same laws as a spin in a magnetic field: it precesses around the direction of the field
- In NMR the *precession frequency* is called the *Larmor frequency*

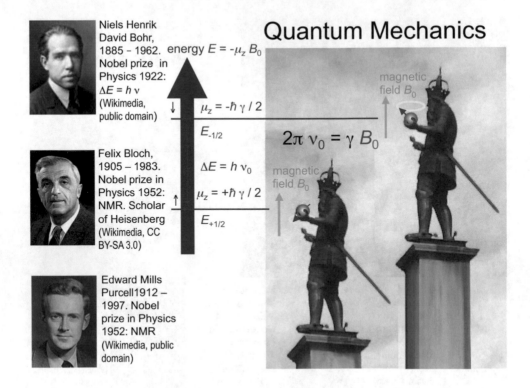

Niels Henrik David Bohr, 1885 – 1962. Nobel prize in Physics 1922: $\Delta E = h\, \nu$ (Wikimedia, public domain)

Felix Bloch, 1905 – 1983. Nobel prize in Physics 1952: NMR. Scholar of Heisenberg (Wikimedia, CC BY-SA 3.0)

Edward Mills Purcell 1912 – 1997. Nobel prize in Physics 1952: NMR (Wikimedia, public domain)

Quantum Mechanics

energy $E = -\mu_z\, B_0$

$\mu_z = -\hbar\, \gamma\, /\, 2$

$E_{-1/2}$

$2\pi\, \nu_0 = \gamma\, B_0$

$\Delta E = h\, \nu_0$

$\mu_z = +\hbar\, \gamma\, /\, 2$

$E_{+1/2}$

magnetic field B_0

magnetic field B_0

Properties of Nuclear Spins

- Following Heisenberg's uncertainty principle, only the component of the spin in the direction of the magnetic field can be determined accurately
- From quantum mechanics it is known that a spin with the *spin quantum number I* can assume $2I + 1$ stable orientations in a magnetic field
- The projection of the spin angular momentum along the direction of the magnetic field is proportional to the *magnetic quantum number m*, where $m = I, I-1, ..., -I$
- $I = \frac{1}{2}$ applies to the nuclei ^1H, ^{13}C, ^{15}N, ^{19}F, ^{31}P, and $I = 1$ to ^2H, ^{14}N
- For nuclei with spin $I = \frac{1}{2}$ there are two possible orientations of it's projection along the axis of the magnetic field: $\uparrow$ ($m = +1/2$) and $\downarrow$ ($m = -1/2$)
- Both orientations differ in the interaction energy $E_m = -\hbar\, \gamma\, m\, B_0$ of the nuclear magnetic dipoles with the magnetic field $\boldsymbol{B}_0$
- According to Bohr's formula $\Delta E = h\, \nu_0$ the energy difference $\Delta E = E_{-1/2} - E_{+1/2} = \hbar\, \gamma\, B_0$ associated with both orientations corresponds to the frequency $\omega_0 = 2\pi\, \nu_0 = \gamma\, B_0$
- Here ν_0 is the precession frequency of the nuclear spins in the magnetic field

Macroscopic Magnetization

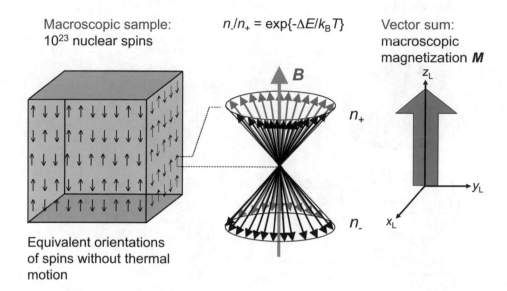

Macroscopic sample:
10^{23} nuclear spins

$$n_-/n_+ = \exp\{-\Delta E/k_B T\}$$

Vector sum:
macroscopic
magnetization M

Equivalent orientations
of spins without thermal
motion

Nuclear Magnetization in Thermodynamic Equilibrium

- At room temperature, all magnetic dipole moments change their orientations in the magnetic field rapidly, because the thermal energy $k_B T$ is orders of magnitude larger than the energy difference $\Delta E = h \nu_0$ between spin states
- The average orientation of all spins in thermal motion is commonly mapped onto the orientations of cold spins not agitated by temperature and aligned either parallel or anti-parallel to the direction of the magnetic field
- All (classical) magnetic dipole moments add as vectors. Their components in each space direction are additive
- The sum of transverse components vanishes
- The sum of longitudinal components forms the *longitudinal magnetization*
- This component is referred to as the *magnetic polarization* of the nuclei or the *nuclear magnetization*
- At room temperature only about 10^{18} spins of 10^{23} spins contribute the macroscopic nuclear magnetization of the sample
- In thermodynamic equilibrium, the nuclear magnetization is oriented parallel to the direction of the magnetic field
- By convention the direction of the magnetic field defines the z direction of the *laboratory coordinate frame* LCF with coordinates x_L, y_L, z_L

Precession of Nuclear Magnetization

Spinning top in a
gravitational field

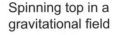

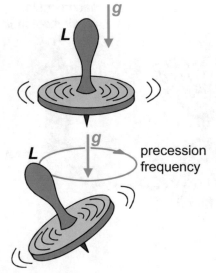

Macroscopic nuclear
magnetization

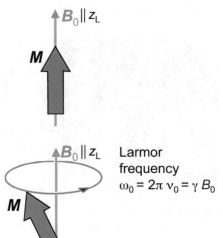

precession
frequency

Larmor
frequency
$\omega_0 = 2\pi\,\nu_0 = \gamma\,B_0$

Bloch's Equation

- When the magnetization M is not aligned with the field direction z_L, it precesses around z_L with the frequency ν_0 in analogy with the precession of a top spinning with angular momentum L exposed to a gravitational force mg
- The precession is described by the equation for the magnetic *spinning top*:

$$\frac{\mathrm{d}}{\mathrm{d}t}\,M = \gamma\,M \times B$$

- This equation states that any change dM of the magnetization M is perpendicular to M and B; therefore M precesses around B
- In general any macroscopic precessional motion is attenuated. This is why Felix Bloch introduced phenomenological attenuation terms:

$$\mathbf{R} = \begin{pmatrix} 1/T_2 & 0 & 0 \\ 0 & 1/T_2 & 0 \\ 0 & 0 & 1/T_1 \end{pmatrix}$$

- The resultant equation is the *Bloch equation*,

$$\frac{\mathrm{d}}{\mathrm{d}t}\,M = \gamma\,M \times B - \mathbf{R}\,(M - M_0),$$

where M_0: initial magnetization, T_1: *longitudinal relaxation time*, and T_2: *transverse relaxation time*, B: any magnetic field
- Note: The Bloch equation formulates a left-handed *rotation* of the *transverse magnetization*. But for convenience sake a right-handed one is used in the illustrations in this book and much of the literature

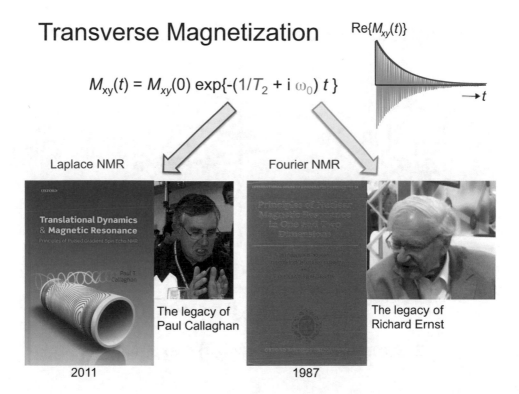

Transverse Magnetization

Re$\{M_{xy}(t)\}$

$$M_{xy}(t) = M_{xy}(0) \exp\{-(1/T_2 + i\,\omega_0)\,t\}$$

$\longrightarrow t$

Laplace NMR

Fourier NMR

The legacy of
Paul Callaghan

The legacy of
Richard Ernst

2011

1987

Transverse Magnetization

- Bloch's equation can be solved when spelling it out for the magnetization components with the *thermodynamic equilibrium magnetization* M_0,

  $dM_{xL}/dt = \gamma\,(M_{yL}\,B_{zL} - M_{zL}\,B_{yL}) - M_{xL}/T_2$

  $dM_{yL}/dt = \gamma\,(M_{zL}\,B_{xL} - M_{xL}\,B_{zL}) - M_{yL}/T_2$

  $dM_{zL}/dt = \gamma\,(M_{xL}\,B_{yL} - M_{yL}\,B_{xL}) - (M_{zL} - M_0)/T_1$

- In the laboratory coordinate frame, the magnetic field vector is

  $\boldsymbol{B} = (B_{xL}, B_{yL}, B_{zL})^\dagger = (2B_1\cos\omega_{TX}t, 0, B_0)^\dagger$,

  where ω_{TX} is the transmitter frequency

- Writing the transverse magnetization in complex form as $M_{xy} = M_{xL} + iM_{yL}$, for $B_1 = 0$ the evolution of the transverse magnetization follows the equation

  $dM_{xy}/dt = -\,i\,\gamma\,M_{xy}\,B_0 - M_{xy}/T_2$

- With $\omega_0 = \gamma\,B_0$, the solution provides the *free induction decay* (*FID*) in the laboratory frame, describing an attenuated left-handed *rotation*,

  $M_{xy}(t) = M_{xy}(0) \exp\{-(1/T_2 + i\,\omega_0)t\} = M_{xy}(0) \exp\{-t/T_2\} \exp\{-i\,\omega_0 t\}$

- This FID can be generated with an rf excitation impulse. Therefore it is also known as the NMR impulse response
- The FID is the product of a decaying and an oscillating exponential function
- In real matter many decay rates $1/T_2$ and oscillation frequencies ω_0 arise
- *Distributions of frequencies* are retrieved from $M_{xy}(t)$ by *Fourier transformation*. *Distributions of relaxation rates* are retrieved from $|M_{xy}(t)|$ by algorithms reminiscent of the *inverse Laplace transformation*

Magnetic Fields in an Oscillator Circuit

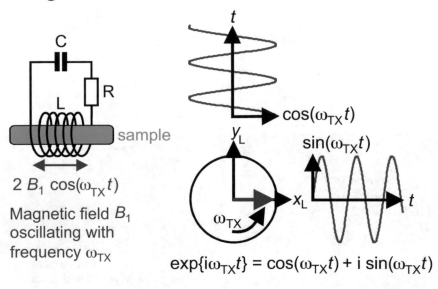

$$\exp\{i\omega_{TX}t\} = \cos(\omega_{TX}t) + i\,\sin(\omega_{TX}t)$$

$$2\,B_1\cos(\omega_{TX}t) = B_1[\exp\{i\omega_{TX}t\} + \exp\{-i\omega_{TX}t\}]$$

Contacting Nuclear Magnetization

- Nuclear magnetization can be rotated away from the direction z_L of the magnetic field $\boldsymbol{B}_0$, which keeps the spins aligned, by a magnetic field $\boldsymbol{B}_1$ perpendicular to $\boldsymbol{B}_0$ oscillating in resonance with the precession frequency
- This field $\boldsymbol{B}_1$ oscillates in the *radio-frequency* (rf) regime with frequency ω_{TX} and is generated by a *transmitter* TX
- For maximum interaction of the rotating field with the nuclear magnetization both frequencies need to match, defining the *resonance condition* $\omega_{TX} = \omega_0$
- Radio-frequency electromagnetic waves are produced by currents oscillating in electronic circuits and are emitted from transmission antennas
- An *electric oscillator* consists of a *coil* with inductance L, a capacitor with capacitance C, and a resistor with resistance R
- The coil generates a linearly polarized, oscillating magnetic field $2B_1\cos(\omega_{TX}t)$
- Two orthogonal, linearly polarized waves $\cos(\omega_{TX}t)$ and $\sin(\omega_{TX}t)$ generate a rotating wave
- The linearly polarized wave $2\cos(\omega_{TX}t)$ of the coil can be decomposed into a right rotating wave $\exp\{i\omega_{TX}t\}$ and a left rotating wave $\exp\{-i\omega_{TX}t\}$, one of which can be adjusted to resonate with the precessing magnetization
- For optimum use of the oscillating magnetic field, the sample to be investigated is placed inside the coil of the electronic oscillator

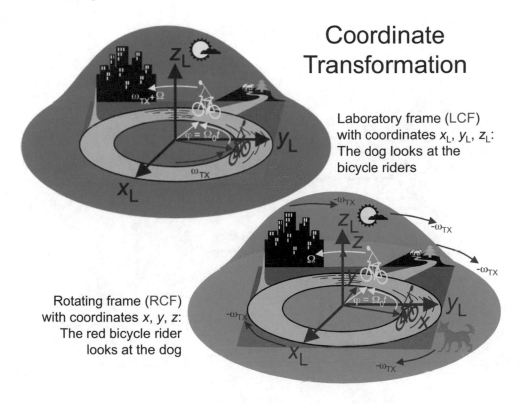

Coordinate Transformation

Laboratory frame (LCF) with coordinates x_L, y_L, z_L: The dog looks at the bicycle riders

Rotating frame (RCF) with coordinates x, y, z: The red bicycle rider looks at the dog

Rotating Coordinate Frame

- Transformations from one coordinate frame into another change the point of view, i. e. they change the mathematics but not the physics
- As the precession of nuclear magnetization is a rotational motion and the rf excitation is a rotating wave, the magnetization is conveniently studied in a *rotating coordinate frame* (RCF) with coordinates x, y, z
- The dog at the traffic circle is positioned in the *laboratory coordinate frame* (LCF) with coordinates x_L, y_L, z_L: For the dog the bicycles are moving in the traffic circle with angular velocities ω_{TX} and $\omega_{TX} + \Omega_0$
- The cyclists on the bicycles are viewing the world from the RCF. They are at rest in their respective RCFs
- For the red cyclist the world in the LCF is rotating against the direction of his bicycle with angular velocity $-\omega_{TX}$
- For the red cyclist the yellow bicycle moves away with angular velocity Ω_0
- The connecting vectors from the center of the traffic circle to the bicycles correspond to the magnetization vectors in the transverse xy plane
- The angular velocity of the RCF as seen in the LCF corresponds to the frequency ω_{TX} of the rf wave
- The spectrometer hardware assures that the NMR signal is measured in the RCF

Action of rf Pulses

Laboratory coordinate frame (LCF)   Rotating coordinate frame (RCF)

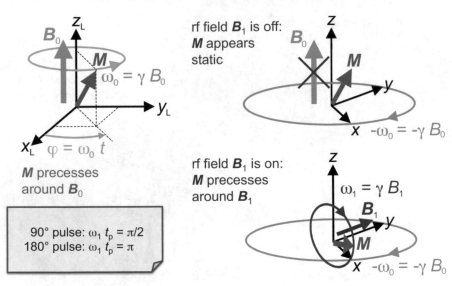

M precesses
around B_0

rf field B_1 is on:
M precesses
around B_1

90° pulse: $\omega_1 t_p = \pi/2$
180° pulse: $\omega_1 t_p = \pi$

Radio-Frequency Pulses

- Radio-frequency (rf) pulses produce a magnetic field B_1, which oscillates with frequency ω_{TX}
- In the rotating coordinate frame RCF, which rotates on resonance with the precession frequency $\omega_{TX} = \omega_0 = \gamma\, B_0$ around the z_L axis, the magnetization M appears at rest even if it is not parallel to the magnetic field B_0
- When the magnetization is not rotating, there is no torque on the magnetization, so that in the RCF rotating on resonance with $\omega_{TX} = \omega_0$ the magnetic field B_0 vanishes
- When the transmitter generates a B_1 field rotating at the frequency ω_{TX} of the RCF in the LCF, this field appears static in the RCF
- In the RCF the magnetization rotates around the B_1 field with frequency $\omega_1 = \gamma\, B_1$ in analogy to the *rotation* with frequency $\omega_0 = \gamma\, B_0$ around the B_0 field in the LCF
- If B_1 is turned on in a pulsed fashion for a time t_p, a 90° pulse is defined for $\omega_1 t_p = \pi/2$ and a 180° pulse for $\omega_1 t_p = \pi$
- The phase φ_{TX} of the *rotating rf field* $B_1 \exp\{i\omega_{TX}t + i\,\varphi_{TX}\}$ defines the direction of the B_1 field in the xy plane of the RCF
- Using this phase the magnetization can be rotated in the RCF around different axes, e. g. $90°_y$ denotes a positive 90° *rotation* around the y axis of the RCF and $180°_x$ a positive 180° rotation around the x axis

Spectrometer Hardware

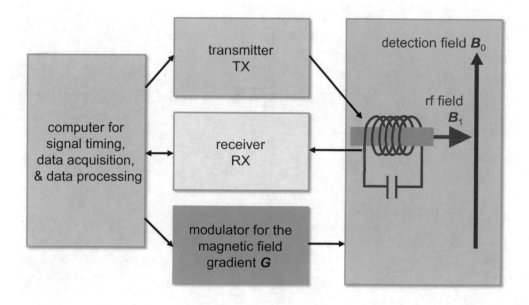

NMR Spectrometer

- The sample is positioned in a magnetic field B_0 inside or next to an rf coil, which is part of an rf *oscillator* tuned to the frequency ω_{TX}
- The oscillator is connected under computer control either to the rf *transmitter* (TX) or to the *receiver* (RX)
- A 90° rf pulse from the transmitter rotates the magnetization from the z_L direction of the B_0 field into the transverse plane
- Following the pulse, the transverse magnetization components precess around the z_L axis of the LCF with frequency ω_0
- According to the dynamo principle, the precessing magnetization induces a voltage in the coil, which oscillates at frequency ω_0
- In the receiver, this signal is mixed with a reference wave at frequency ω_{TX}, and the audio signal at the difference frequency is acquired for further analysis
- This step is the transition into the *rotating coordinate frame*
- Depending on the phase $\varphi_{RX} = 0°$ and 90° of the reference wave $\cos(\omega_{TX}t + \varphi_{RX})$ the *components* $\cos(\omega_0 - \omega_{TX})t$ and $\sin(\omega_0 - \omega_{TX})t$ of the transverse magnetization are measured in the RCF at frequency $\Omega_0 = \omega_0 - \omega_{TX}$
- Usually the in-phase and the *quadrature component* are measured together
- For imaging and flow measurements the spectrometer is equipped with switchable *gradient fields* in x_L, y_L, and z_L directions of the LCF

Fourier NMR

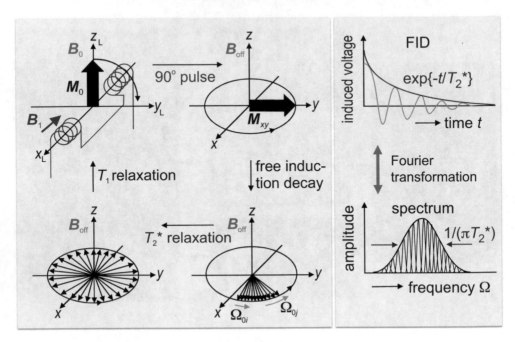

Pulsed Excitation

- Outside a magnetic field the magnetic dipole moments of the nuclear spins are oriented in random directions in space
- In a magnetic field B_0, the nuclear spins partially align with the field and form the *longitudinal magnetization* M_0 parallel to B_0 in a characteristic time T_1 following the rate equation $M_z(t) = M_0 (1 - \exp\{-t/T_1\})$
- T_1 is the *longitudinal relaxation time*
- A resonant 90° rf pulse from the transmitter rotates the magnetization from the z direction of the magnetic field B_0 into the transverse plane of the RCF
- After the pulse the transverse magnetization components $M_{xy,i}$ precess around the z axis of the RCF with the difference frequencies $\Omega_{0i} = \omega_{0i} - \omega_{TX}$
- Each component $M_{xy,i}$ corresponds to a different chemical shift or another position in the sample with a different magnetic field at the site of the nuclei
- The vector sum of the *transverse magnetization* components decays with the time constant T_2^* due to destructive interference of the components with different precession frequencies Ω_{0i}
- T_2^* is the apparent *transverse relaxation time*. It is determined by time-invariant and time-dependent local magnetic fields at the sites of the spins
- The signal-decay envelope is often exponential: $M_{xy}(t) = M_z(0) \exp\{-t/T_2^*\}$
- The signal induced in the coil after an excitation pulse is the FID
- For one component it is $M_{xy}(t) = M_z(0) \exp\{-(1/T_2^* + i\Omega_0)t\}$ after a 90° pulse
- A frequency analysis of the FID by *Fourier transformation* produces the *NMR spectrum* with a *linewidth* $\Delta\Omega = 1/(\pi T_2^*)$

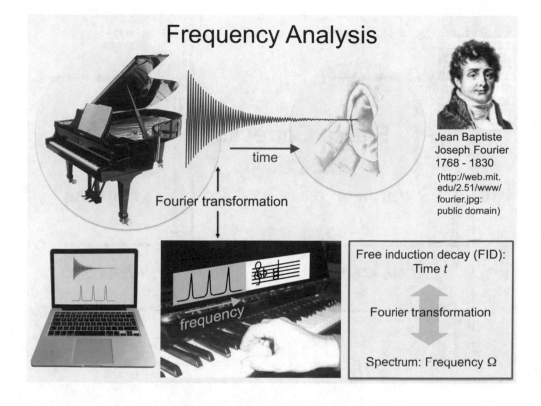

Fourier Transformation

- J.B.J. Fourier introduced the transformation named after him when studying thermal conductivity
- The *Fourier transformation* (FT) is a decomposition of a function $s(t)$ into harmonic waves $\exp\{i\ \Omega t\} = \cos(\Omega t) + i \sin(\Omega t)$ with variable frequency Ω
- In NMR the *FID* $s(t)$ is transformed to the spectrum $S(\Omega)$ of cosine and sine waves: $S(\Omega) = \int s(t) \exp\{+i\Omega t\}\ dt$
- The *spectrum* $S(\Omega) = U(\Omega) + i\ V(\Omega)$ consists of a real part $U(\Omega)$ and an imaginary part $V(\Omega)$
- Often only the magnitude spectrum $|S(\Omega)| = [U(\Omega)^2 + V(\Omega)^2]^{1/2}$ is considered
- The Fourier transformation corresponds to the transformation of an acoustic signal into the pitch and volume of sound
- For the discrete Fourier transformation there is a fast algorithm, which was rediscovered in 1965 by J. W. Cooley and J. W. Tukey
- The algorithm requires discrete representations of time $t = n\ \Delta t$ and frequency $\Omega = n\ \Delta\Omega$ at intervals Δt and $\Delta\Omega$, whereby range and step size are related via the acquisition time $n_{max}\ \Delta t$ of the NMR signal following $\Delta\Omega = 1/(n_{max}\ \Delta t)$
- The abscissa of the discrete spectrum corresponds to the keys of a piano
- The spectral amplitude corresponds to the volume of a given tone
- NMR spectroscopy and imaging with pulsed excitation rely on the Fourier transformation for data processing. They are referred to as *Fourier NMR*
- The product of two Fourier conjugated variables, e.g. t and Ω, is always an angle. It is referred to as *phase* φ

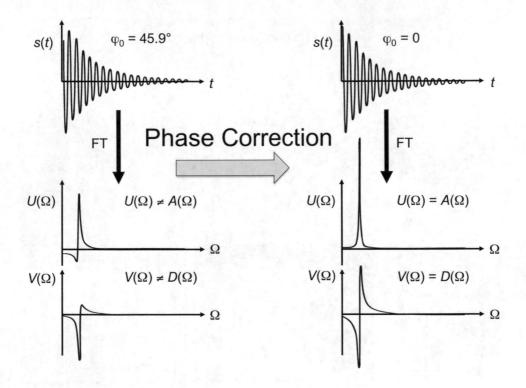

Signal Processing

- Depending on the electronics of the receiver, the NMR signal
  $s(t) = s(0) \exp\{-[1/T_2 + i\Omega_0] t + i \varphi_0\}$ of a magnetization component is
  recorded in practice with a phase offset φ_0
- For $\varphi_0 = 0$ the real part $U(\Omega)$ of the *Fourier transform* $S(\Omega)$ is an *absorption
  signal* $A(\Omega)$ and the imaginary part $V(\Omega)$ a *dispersion signal* $D(\Omega)$
- For $\varphi_0 \neq 0$ the real and imaginary components of $S(\Omega)$ are mixed in $U(\Omega)$ and
  $V(\Omega)$, and the associated complex *spectrum* $S(\Omega) = U(\Omega) + i\,V(\Omega) = [A(\Omega) +
  i\,D(\Omega)] \exp\{i\,\varphi_0\}$ has to be corrected in phase by multiplication with $\exp\{-i\,\varphi_0\}$
- The correction phase φ_0 consists of a frequency-dependent and a
  frequency-independent part
- The frequency-independent part can be adjusted by software before data
  acquisition via the receiver reference phase φ_{RX}
- The frequency dependent part is determined by the time the signal takes to
  pass through the spectrometer and by the receiver *deadtime* following an
  excitation pulse
- For optimum resolution the spectrum is needed in pure absorptive mode as
  $A(\Omega)$
- A frequency-dependent *phase correction* of the spectrum is a routine data
  processing step in high-resolution NMR spectroscopy

RF Excitation and Effective Field

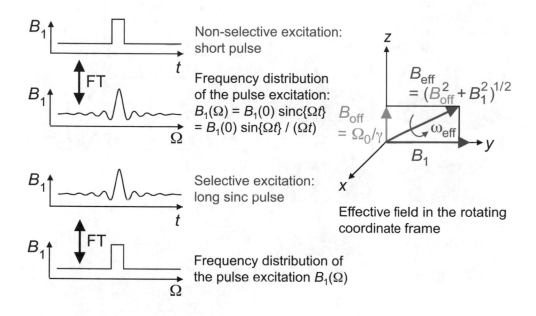

Non-selective excitation: short pulse

Frequency distribution of the pulse excitation:
$B_1(\Omega) = B_1(0) \, \text{sinc}\{\Omega t\}$
$= B_1(0) \sin\{\Omega t\} / (\Omega t)$

Selective excitation: long sinc pulse

Frequency distribution of the pulse excitation $B_1(\Omega)$

$$B_{\text{eff}} = (B_{\text{off}}^2 + B_1^2)^{1/2}$$

$$B_{\text{off}} = \Omega_0/\gamma$$

Effective field in the rotating coordinate frame

Frequency Distributions

- The *rotating coordinate frame* RCF rotates with the rf frequency ω_{TX}
- Magnetization components M_i rotate in the RCF with frequencies $\Omega_{0i} = \omega_{0i} - \omega_{\text{TX}}$
- To contact magnetization components in a range of frequencies Ω_{0i} with an *rf pulse* at frequency ω_{TX}, the pulse has to have a *frequency bandwidth*
- This frequency range is approximated by the inverse of the pulse width t_p
- A better measure for the bandwidth of the excitation is the *Fourier transform* of the excitation pulse
- For a rectangular pulse this is the *sinc* function with wiggles right and left
- Vice versa, the excitation can be made sharply frequency selective by excitation with an rf pulse having a sinc shape in the time domain
- This simple Fourier relationship is a convenient approximation valid only for flip angles smaller than 30°
- For a given component the offset frequency Ω_0 corresponds to a magnetic *off-set field* $B_{\text{off}} = \Omega_0/\gamma$ along the z axis of the RCF
- The magnetization always rotates around the *effective field* B_{eff}, which is the vector sum of the offset field and the rf field with components Ω_0/γ and B_1
- The *rotation angle* around the effective field is given by $\gamma \, B_{\text{eff}} \, t_p = \omega_{\text{eff}} \, t_p$
- The rotation axis is close to the *xy* plane if $|B_1| \gg |\Omega_0/\gamma|$
- If $|B_1| \ll |\Omega_0/\gamma|$, longitudinal magnetization cannot be rotated into the *xy* plane
- Low B_1 amplitudes are employed for frequency-selective excitation

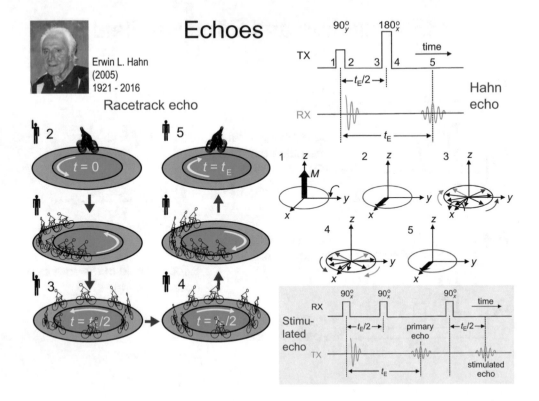

Erwin L. Hahn
(2005)
1921 - 2016

Relaxation

- *Relaxation* denotes the loss of *transverse magnetization* and build-up of *longitudinal magnetization*. The respective time constants are T_2 and T_1
- The loss of transverse magnetization due to different time-invariant local magnetic fields can stroboscopically be reversed by formation of *echoes*
- To generate a *racetrack echo* all bicyclists start at the same time but ride with different speeds. After a certain time all turn back and meet at the starting line forming the *echo* after twice that time
- Their total riding time is the *echo time* t_E
- The NMR *echo* has accidentally been discovered in 1949 by Erwin Hahn
- To form a *Hahn echo* all transverse magnetization components are rotated by 180° around an axis in the *xy* plane
- The direction of precession is maintained with this change of positions on the circle, and all magnetization components refocus at time t_E
- If some components randomly change their precession frequencies, the *echo amplitude* is irreversibly reduced
- Random frequency changes arise from fluctuating local magnetic fields associated with molecules in motion
- Transverse relaxation denotes the irreversible loss of the echo amplitude
- Both *relaxation times* T_1 and T_2 are determined by the type and time scale of molecular motion
- When splitting the 180° pulse of the Hahn-echo sequence into two 90° pulses separated by a time delay, one obtains the *stimulated echo* sequence

Echoes and Inhomogeneous Magnetic Fields

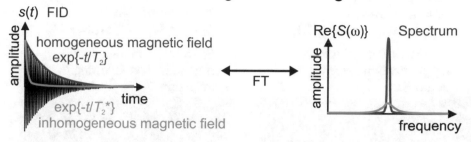

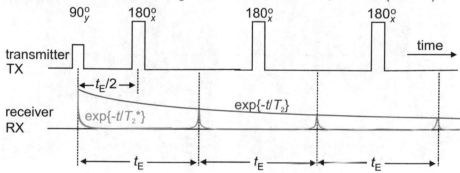

Multiple Echoes

- Transverse relaxation is often exponential with the time constant T_2
- In *inhomogeneous magnetic fields*, the FID decays faster with $T_2^* < T_2$
- The resonance signal in inhomogeneous magnetic fields is broad and small
- The envelope of the *FID* in homogeneous fields can be observed stroboscopically in inhomogeneous fields via the amplitude of many time-shifted echoes
- Instead of many *Hahn echoes* with different echo times t_E, the echo envelope can be observed by a single train of multiple Hahn echoes
- The rf pulse scheme for excitation of multiple Hahn echoes is the *CPMG sequence* named after their discoverers Carr, Purcell, Meiboom, and Gill
- The repetition times of $5T_1$ for regeneration of longitudinal magnetization between generation of different Hahn echoes are eliminated
- A waiting time of $5T_1$ is needed to regain about 99% of the *thermodynamic equilibrium magnetization*, because $\exp\{-5\} = 0.007$
- Besides the Hahn echo, stimulated echo, and CPMG echo train there are many other echoes and multiple-echo schemes to partially recover signal loss due to different nuclear spin interactions
- In the Hahn echo maximum, the inhomogeneity of the $\boldsymbol{B}_0$ field and the spread in chemical shifts do not affect the NMR signal

Pulse Sequences for Measurement of T_1

Saturation recovery:
$M_z(t_0) = M_0(1 - \exp\{-t_0/T_1\})$

Build-up of longitudinal
magnetization following
saturation

Inversion recovery:
$M_z(t_0) = M_0(1 - 2\exp\{-t_0/T_1\})$

Build-up of longitudinal
magnetization following
inversion

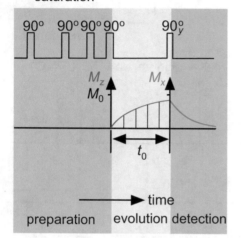 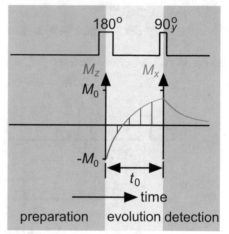

Determination of T_1

- *Longitudinal magnetization* cannot be directly observed
- Its momentary value can be interrogated via the amplitude of the FID following a 90° pulse
- There are two methods to measure the build-up of longitudinal magnetization, by *saturation recovery* and by *inversion recovery*
- For *saturation*, the spin system is irradiated with an aperiodic sequence of 90° pulses, which destroys all magnetization components
- For *inversion* of the longitudinal magnetization a 180° pulse is applied following the establishment of equilibrium magnetization after a waiting time of $5T_1$
- After such preparation of the initial magnetization a variable evolution time t_0 follows for partial recovery of the *thermodynamic equilibrium magnetization*
- After the waiting time t_0, the momentary value of the longitudinal magnetization is converted into the amplitude of the *transverse magnetization* by a 90° pulse
- The amplitude of the transverse magnetization is measured and evaluated for different values of t_0
- In homogeneous spin systems, the longitudinal relaxation follows an exponential law

Probing Resonance

Forced oscillation:
One resonance
probed at a time

Free oscillation:
Many resonances
probed at a time.
All initially in phase

Stochastic resonance:
Many resonances probed
at random and out of
phase

⟶ time

Methods for Measuring Resonances

- NMR is a *resonance phenomenon*. Consequently the basic NMR methods are the ones known from elementary physics to probe resonance phenomena
- There are three fundamental methods: forced oscillation, free oscillation and stochastic resonance
- *Forced oscillations* are known from wind instruments and string instruments like the violin, which are continuously agitated to produce the sound
- *Free oscillations* are known from string instruments like the piano or the guitar where strings are set into vibration with impulses
- *Stochastic resonance* is known from spectroscopy with the Michelson interferometer, where white noise randomly triggers vibrations, and the cross-correlation of excitation and response retrieves the impulse response
- The most versatile method of all is pulsed excitation. This already becomes clear when comparing sheet music for the piano and the violin

Pulse sequence for piano

Forced oscillations for violin

CW-, Fourier, and Stochastic NMR

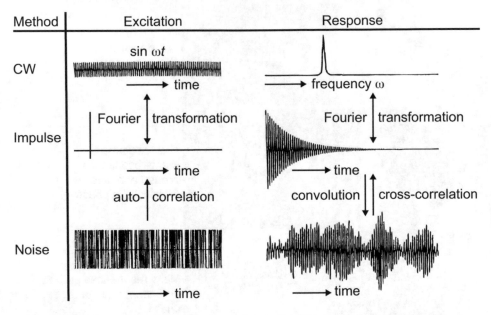

From B. Blümich, **Prog. Nucl. Magn. Reson. Spectr. 19** (1987) 331, Figs. 3, 5 with permission

Basic NMR Methods

- Probing resonance with *pulsed excitation* and measuring the impulse response or free oscillation is the state of the art when measuring NMR data
- With pulsed excitation, the acquired signal can be conditioned by manipulating the initial magnetization in preparation and evolution periods preceding the acquisition period
- Pulsed NMR is uniquely suited for extension to *multi-dimensional NMR*
- With short pulses large spectral widths can be excited, and many frequency components can be simultaneously measured (*multiplex advantage*)
- Measuring resonance by *forced oscillations* with *continuous waves* (CW) is slow, because the excitation frequency needs to be swept across the frequency range of the spectrum
- With *noise excitation* (stochastic NMR) large bandwidths are excited and can be measured simultaneously, but the experiment cannot readily be divided into different periods such as preparation, evolution, and detection periods
- The *excitation power* in CW NMR and stochastic NMR is several orders of magnitude lower than in pulsed NMR
- The different measuring methods are related via Fourier transformation, auto-, and cross-correlation of excitation and response

Space Encoding

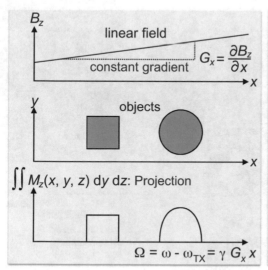 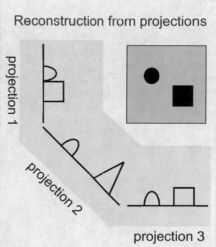

Magnetic Resonance Imaging (MRI)

- By exploring the proportionality of the resonance frequency ω_0 and the applied magnetic field B, signals from different positions in the sample can be discriminated if the magnetic field changes with position
- For a linear change of the magnetic field B_z with position, the magnetic field is characterized by a constant (space-invariant) *gradient G*
- In a linear field, the resonance frequency ω_0 is proportional to position, and the frequency axis of the NMR spectrum can be replaced by a space axis if the spectrum contains only one line
- The signal amplitude is determined by the number of nuclear spins at a particular position along the gradient direction
- This number is obtained by summation over all nuclei in the other two space directions
- Due to the large number of nuclei, the sum can be replaced by an integral over the magnetization density $M_z(x,y,z)$
- This integral is called a *projection*. It is a 1D image of the object
- $M_z(x,y,z)$ is also referred to as *spin density*
- From a set of projections acquired for different gradient directions an image of the object can be reconstructed in analogy to X-ray computed tomography
- The first magnetic resonance images were obtained in this way

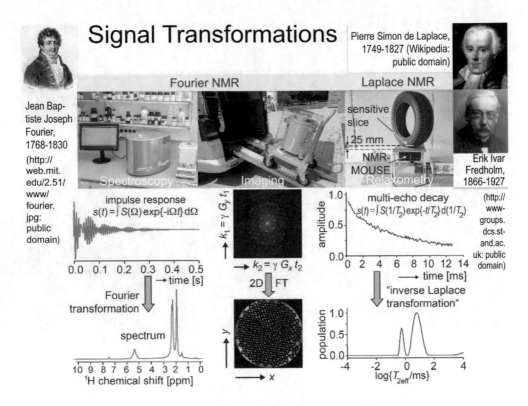

Fourier and Laplace NMR

- Bloch's equation expresses the FID of a magnetization component in the rotating frame as the product of a real and a complex exponential function
$$s(t) = M_{xy}(t) = M_{xy}(0) \exp\{-(1/T_2 + i\Omega_0)t\} = M_{xy}(0) \exp\{-(t/T_2\} \exp\{-i\Omega_0 t\}$$
- The *complex signal* $s(t)$ is Fourier transformed into a *spectrum* $S(\Omega)$ or an *image* revealing the *distribution of amplitudes* versus frequency or space
- The real exponential enters the linewidth in the spectrum or the spatial resolution in the image
- In inhomogeneous fields chemical shift differences cannot be observed (equivalent to $\Omega_0 = 0$), but the relaxation signal $s(n) = M_{xy}(0) \exp\{-(n\, t_E/T_2\}$ from the real exponential can be recalled in the maxima of a multi-echo train
- A relaxation signal $s(t)$ is inverted to a *distribution of relaxation rates* $S(1/T_2)$ by inverting a *Fredholm equation* of the first kind. In the NMR community this is somewhat generously referred to as *inverse Laplace transformation*
- The inversion algorithm assumes the relaxation signal to be a sum of real-valued exponential functions similar to the Fourier transformation, which takes the signal to be a sum of complex-valued exponential functions
- Forward and inverse Fourier transformations are stable operations even in the presence of noise, but Laplace inversion is not and needs to be stabilized
- *Fourier NMR* refers to the complex-valued exponential of the FID and *Laplace NMR* to the real-valued exponential

Analyzing Multi-Modal Distributions

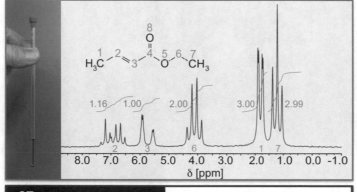

43 MHz ^{1}H NMR spectrum of ethyl-crotonate. The relative multiplet integrals reveal the numbers of protons in each chemical group

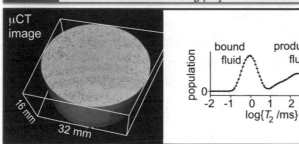

T_2 distribution of water in travertine sandstone. The peak integrals are proportional to the amounts of bound and producible fluid

Component Analysis

- Most samples contain many magnetization components with different relaxation times T_{2i} and resonance frequencies Ω_{0i}, and their magnetization sum is measured in terms of the detected signal $s(t)$
$$s(t) = M_{xy}(t) = \Sigma_i\, M_{xyi}(0)\, \exp\{-(1/T_{2i} + i\Omega_{0i})t\}$$
- *Distributions of frequencies* Ω_{0i} (NMR spectra) and *distributions of relaxation rates* $1/T_{2i}$ with $\Omega_{0i} = 0$ are uniquely suited to quantify component fractions in terms of spin fractions $x_i = {}_{\Omega_1}\!\int^{\Omega_2} S(\Omega)\, d\Omega\, /\, s(0)$ and $x_i = {}_{(1/T_2)_1}\!\int^{(1/T_2)_2} S(1/T_2)$ $d(1/T_2)\, /\, s(0)$ by integration of peaks in the distributions $S(\Omega)$ and $S(1/T_2)$ obtained by Fourier and inverse Laplace transformation of $s(t)$, respectively
- *Peak integrals* from *a high-resolution NMR spectrum* provide relative concentrations of chemical groups in a molecule or different molecules in a mixture
- *Peak integrals* from *distributions of relaxation times* provide relative concentrations of physical structures such as the fractions of amorphous and interfacial domains in *semi-crystalline polymers* and the relative amounts of *bound fluid* and *producible fluid* in porous medium
- If peaks or groups of peaks overlap, then two- and *multi-dimensional NMR* techniques may help to identify and separate the peaks belonging to different groups such as the *multiplets* in *J*-coupled ^{1}H spectra and the component spectra from mixtures
- For example 2D MRI separates a 1D projection into a 2D image

3. Spectroscopy

The dipole-dipole interaction
Interaction anisotropy
Further spin interactions
The density matrix
Solid-state NMR spectroscopy
Multi-quantum NMR
Multi-dimensional NMR spectroscopy

© Springer Nature Switzerland AG 2019
B. Blümich, *Essential NMR*,
https://doi.org/10.1007/978-3-030-10704-8_3

General Formalism

$$\mu = \gamma\,p = \gamma\,\hbar\,I$$

$$E = I^{\dagger}\,\mathbf{P}\,\mathbf{V}$$

μ: nuclear magnetic dipole moment
γ: gyro-magnetic ratio
p: vector of the angular momentum
$\hbar = h/2\pi$, h: Planck's constant
I: nuclear spin vector operator

E: interaction energy
I: nuclear spin vector operator
$\mathbf{P}$: coupling tensor
$\mathbf{V}$: coupling vector partner

$$E = \begin{bmatrix} I_x & I_y & I_z \end{bmatrix} \begin{bmatrix} P_{xx} & P_{xy} & P_{xz} \\ P_{yx} & P_{yy} & P_{yz} \\ P_{zx} & P_{zy} & P_{zz} \end{bmatrix} \begin{bmatrix} V_x \\ V_y \\ V_z \end{bmatrix}$$

$\nu_{1H} = \Delta E/h$	Spin interaction	Spin partner
100 MHz	Zeeman interaction	B_0
10 kHz	chemical shift	$-\sigma B_0$
100 kHz	rf excitation	B_1
10 MHz	quadrupole coupling	I
50 kHz	dipole-dipole interaction	I'
5 Hz	indirect coupling	I'

Interactions Between Spins

- The magnetic *dipole moment* μ of a nucleus is proportional to its spin I
- The dipole moment and the spin are vectors with magnitude and orientation
- Also, the magnetic field B is a vector quantity
- The magnetic moments interact with each other and with the magnetic field
- The strength of an interaction is measured by the *interaction energy E*. This is a quantity without orientation. It is a scalar
- An interaction is formally described by the product of two quantities
- For the product of the spin vector I and its coupling partner to be a scalar, the coupling partner must be a vector V
- The coupling partners of a spin I are further spins I' and the magnetic field B composed of B_0, B_1, as well as the field induced by the shielding electrons
- In the simplest case, the interaction is described by the scalar product of two vectors, for example, by $E \propto I^{\dagger}\,V$, where $\dagger$ denotes the transpose
- To describe orientation-dependent interactions, a *coupling tensor* $\mathbf{P}$ must be introduced, so that $E = I^{\dagger}\,\mathbf{P}\,V$
- The significance of $\mathbf{P}$ is elaborated below by example of the *dipole-dipole interaction*
- Interactions of a spin with a magnetic field B are distinguished from interactions of one spin with another spin. The interactions between two spins formally include the nuclear *quadrupole interaction*

Dipole-Dipole Interaction

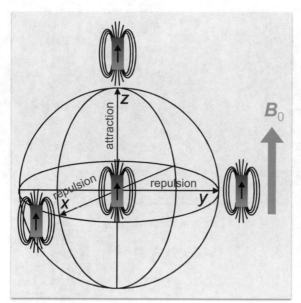 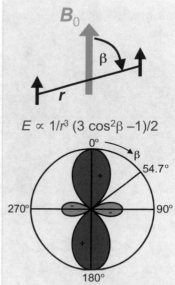

$$E \propto 1/r^3 \, (3 \cos^2\beta - 1)/2$$

Two Interacting Dipoles

- Interactions of nuclear spins are interactions of elementary quantities for which the laws of quantum mechanics apply
- The classical treatment of nuclear interactions is at best an approximation which provides some intuitive insights
- The orientation dependence of the interaction energy can be understood by considering two classical bar magnets or compass needles
- For parallel orientation, the magnets repel each other when they are side by side, and they attract each other when one is on top the other
- Viewing the dipole as a distortion of a sphere, the *orientation dependence* of the *dipole-dipole interaction* can be described by the difference between an isotropic sphere and a simply deformed sphere, i.e. a rotational ellipsoid
- This difference is quantified by the *second Legendre polynomial* $P_2(\cos\beta) = (3 \cos^2\beta - 1)/2$
- The angle β denotes the angle between the magnetic field vector $\boldsymbol{B}_0$ and the vector $\boldsymbol{r}$ which connects the two point dipoles
- General deformations are described by the *spherical harmonic functions*, which also describe the electron orbitals of the hydrogen atom

Deformation of a Circle and a Sphere

Deformation of a circle

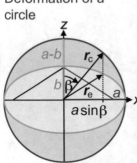

Circle: $r_c = \begin{pmatrix} a\sin\beta \\ 0 \\ a\cos\beta \end{pmatrix}$ Ellipse: $r_e = \begin{pmatrix} a\sin\beta \\ 0 \\ b\cos\beta \end{pmatrix}$

Deviation of an ellipse from
a circle in z direction: $(r_c - r_e)^2 = (a-b)^2\cos^2\beta$

Subtraction of the
mean along z: $P_{aniso} = \dfrac{1}{3}(a-b)^2(3\cos^2\beta - 1)$

Normalization of the angle-
dependent part along z: $P_{aniso} = \dfrac{2}{3}(a-b)^2\dfrac{1}{2}(3\cos^2\beta - 1)$

$\beta = 0°$: principal value $\delta = (2/3)(a-b)^2$
      anisotropy parameter $\Delta = (a-b)^2$

Axially symmetric deformation
of a sphere

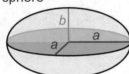

Deformation ellipsoid

$P_{anisoZZ} = \dfrac{2}{3}(a-b)^2$

$P_{anisoXX} = \dfrac{-1}{3}(a-b)^2$

$P_{anisoYY} = \dfrac{-1}{3}(a-b)^2$

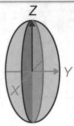

Second Legendre Polynomial

- The *second Legendre polynomial* P_2 describes the geometrically simplest deformation of a circle
- It quantifies the quadratic deviation of a circle from an *ellipse*, where both figures are generated by a thread of length $2r$, where r corresponding to the diameter of the circle and the long axis of the ellipse
- The difference between the circle and the ellipse is only in the direction of the small axis of the ellipse
- The average of the quadratic deviation is subtracted from that deviation to obtain a function with a zero mean value
- The resultant function is normalized to 1 for the angle $\beta = 0°$
- The result is proportional to the second Legendre polynomial
  $P_2 = (3\cos^2\beta - 1)/2$
- The *principal value* is obtained for the angle $0°$. Its value amounts to $2/3$ of the *anisotropy parameter*
- The values of P_2 for $0°$ and $90°$ can be defined as the half axes $P_{anisoZZ}$ and $P_{anisoXX} = P_{anisoYY}$ of another *rotational ellipsoid*
- Without transverse symmetry $P_{anisoXX} \neq P_{anisoYY}$, so that an *asymmetry parameter* can be defined as $\eta = (P_{anisoYY} - P_{anisoXX})/\delta$

Orientation Dependence

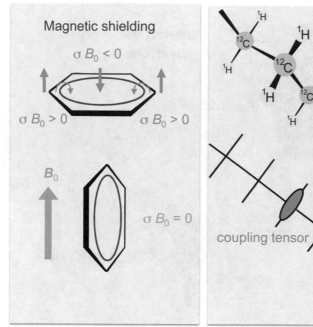

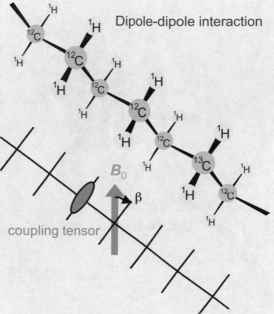

Anisotropy of the Interaction

- The ellipsoid defined by $P_{anisoXX}$, $P_{anisoYY}$, and $P_{anisoZZ}$ describes the *interaction anisotropy* in the limit of coupling energies weak compared to the spin interaction with the magnetic field $\boldsymbol{B}_0$ (*Zeeman interaction*)
- Upon *rotation* of the molecules, the spins stay aligned with the $\boldsymbol{B}_0$ field
- For the *dipole-dipole interaction* $P_{anisoXX} = P_{anisoYY}$, but for the *chemical (magnetic) shielding* and the *quadrupole interaction* $P_{anisoXX} \neq P_{anisoYY}$
- The interactions of a spin with the magnetic fields $\boldsymbol{B}_0$ and $\boldsymbol{B}_1$ are isotropic, i.e. in this case $P_{anisoXX} = P_{anisoYY} = P_{anisoZZ} = 0$
- For *anisotropic interactions* the orientation of the interaction ellipsoid within the molecule is determined by the chemical structure
- In case of the dipole-dipole interaction, the principal (long) axis of the interaction ellipsoid is aligned along the direction of the inter-nuclear vector
- Also the chemical shielding is anisotropic. Here, the orientation of the interaction ellipsoid in the molecular frame can be obtained by means of quantum-mechanical calculations of the electron orbitals
- To describe the interaction ellipsoid, three values suffice in the coordinate frame of the ellipsoid. These are the *eigenvalues* of the interaction tensor
- In another coordinate frame, the molecular frame or the LCF, for example, the orientation of the interaction ellipsoid has to be specified as well. There one needs 6 values to describe the interaction tensor

Anisotropy and Asymmetry

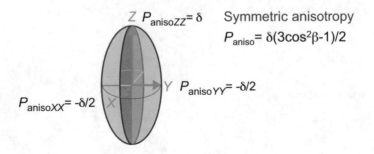

$Z\ P_{anisoZZ}= \delta$ Symmetric anisotropy
$P_{aniso}= \delta(3\cos^2\beta-1)/2$

$Y\ P_{anisoYY}= -\delta/2$

$P_{anisoXX}= -\delta/2$ X

$Z\ P_{anisoZZ}= \delta$

Asymmetric anisotropy
$P_{aniso}= \delta\ [3\cos^2\beta-1-\eta\ \sin^2\beta\ \cos(2\alpha)]/2$

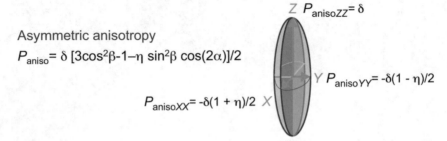

$Y\ P_{anisoYY}= -\delta(1 - \eta)/2$

$P_{anisoXX}= -\delta(1 + \eta)/2\ X$

Anisotropic and Asymmetric Couplings

- The *interaction anisotropy* described by the *second Legendre polynomial*
  $P_2 = (3 \cos^2\beta -1)/2$ is represented by a rotational ellipsoid with the half axis
  $P_{anisoZZ}$ for $\beta = 0°$ and the half axes $P_{anisoXX} = P_{anisoYY}$ for $\beta = 90°$
- For an *asymmetry* in the transverse plane $P_{anisoXX} \neq P_{anisoYY}$
- To describe this asymmetry in spherical coordinates, another angle α needs
  to be introduced
- The ellipsoid which, in this case, describes the interaction, has the shape of
  an American football pressed flat, $P_{ansio} = \delta[3 \cos^2\beta -1 - \eta \sin^2\beta \cos(2\alpha)]/2$
- It possesses the half axes $P_{anisoZZ} = \delta$, $P_{anisoXX} = -\delta(1 + \eta)/2$, and
  $P_{anisoYY} = -\delta(1 - \eta)/2$
- Examples for *asymmetric spin couplings* are the *chemical shielding* and the
  electric quadrupole interaction
- In liquids, the angles α and β change rapidly and isotropically in a random
  fashion. On average the anisotropy of the interaction vanishes over time
- For *symmetric interactions* like the *dipole-dipole interaction* $\eta = 0$, and the
  anisotropic part of the interaction vanishes at the *magic angle*
  $\beta_m = arcos\{(1/3)^{1/2}\} = 54.7°$

Rotation of Tensors

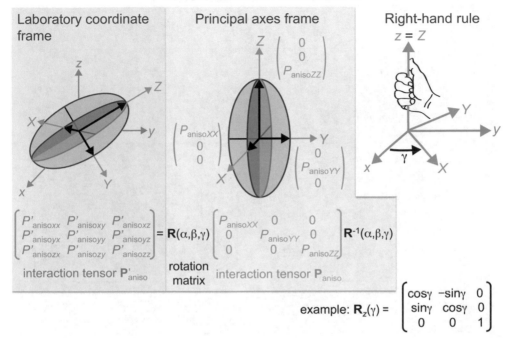

Laboratory coordinate frame

Principal axes frame

Right-hand rule

interaction tensor $\mathbf{P}'_{aniso}$

rotation matrix

interaction tensor $\mathbf{P}_{aniso}$

example: $\mathbf{R}_z(\gamma) = \begin{bmatrix} \cos\gamma & -\sin\gamma & 0 \\ \sin\gamma & \cos\gamma & 0 \\ 0 & 0 & 1 \end{bmatrix}$

Vectors, Matrices, and Tensors

- The half axes of the *interaction ellipsoid* define orthogonal vectors in a Cartesian coordinate frame, which is called the *principal axes frame*
- These three vectors are arranged to form a 3×3 matrix
- A matrix with physical significance is called a *tensor*
- In the principal axes frame, the interaction tensor is diagonal
- The numbers on the diagonal of this tensor are called *eigenvalues*
- In any other coordinate frame, the interaction tensor is not diagonal and appears rotated away from its orientation in the principal axes frame
- From there the tensor can be returned to diagonal form by a counter rotation
- A vector r is rotated by a *rotation matrix* $\mathbf{R}(\gamma)$ according to $r' = \mathbf{R}(\gamma)\,r$
- A matrix or a tensor $\mathbf{P}_{aniso}$ is rotated according to $\mathbf{P}'_{aniso} = \mathbf{R}(\gamma)\,\mathbf{P}_{aniso}\,\mathbf{R}^{-1}(\gamma)$
- For example, with $r'' = \mathbf{R}\,\mathbf{P}_{aniso}\,r = \mathbf{R}\,\mathbf{P}_{aniso}\,\mathbf{R}^{-1}\mathbf{R}\,r = \mathbf{P}'_{aniso}\,r'$, where $\mathbf{P}'_{aniso} = \mathbf{R}\,\mathbf{P}_{aniso}\,\mathbf{R}^{-1}$ and $r' = \mathbf{R}\,r$ are valid in the rotated frame
- A rotation matrix is specified by the rotation axis and the *rotation angle*. For example, $\mathbf{R}_z(\gamma)$ describes a rotation around the z axis by the angle γ
- Any *rotation* around an arbitrary axis is described by three rotations around orthogonal axes through the three *Euler angles* α, β, γ
- One successively performs the rotations $\mathbf{R}_z(\alpha)$, $\mathbf{R}_x(\beta)$, $\mathbf{R}_z(\gamma)$
- The row vectors of the rotation matrix, which diagonalizes the matrix $\mathbf{P}'_{aniso}$ are called *eigenvectors* of the matrix $\mathbf{P}'_{aniso}$
- The eigenvectors are the unit vectors in the directions of the principal axes of the interaction ellipsoid in the rotated frame

Spin Interactions

Interaction	Coupling partner	Isotropic part	Anisotropy parameter Δ	Asymmetry parameter η
Zeeman interaction	B_0	ω_0	0	0
chemical shielding	B_0	$-\gamma\sigma B_0$	✓	✓
rf excitation	B_1	ω_1	0	0
quadrupole interaction	I	0	✓	✓
dipole-dipole interaction	I'	0	✓	0
indirect coupling	I'	J	✓	✓

General interaction tensor: $\mathbf{P} = \qquad \mathbf{P}_{iso} \quad + \qquad \mathbf{P}_{aniso}$

Principal axes frame: $\mathbf{P} = P_{iso} \begin{bmatrix} 1 & 0 & 0 \\ 0 & 1 & 0 \\ 0 & 0 & 1 \end{bmatrix} + \begin{bmatrix} P_{anisoXX} & 0 & 0 \\ 0 & P_{anisoYY} & 0 \\ 0 & 0 & P_{anisoZZ} \end{bmatrix}$

Laboratory coordinate frame: $\mathbf{P}' = P_{iso} \begin{bmatrix} 1 & 0 & 0 \\ 0 & 1 & 0 \\ 0 & 0 & 1 \end{bmatrix} + \begin{bmatrix} P'_{anisoxx} & P'_{anisoxy} & P'_{anisoxz} \\ P'_{anisoyx} & P'_{anisoyy} & P'_{anisoyz} \\ P'_{anisozx} & P'_{anisozy} & P'_{anisozz} \end{bmatrix}$

Rules:

$P'_{anisoXX} + P'_{anisoYY} + P'_{anisoZZ} = \mathrm{Tr}\{\mathbf{P}'_{aniso}\} = 0$ anisotropy: $\Delta = P'_{zz} - (P'_{xx} + P'_{yy})/2 = 3\delta/2$

$\mathrm{Tr}\{\mathbf{P}'\} = 3 P_{iso}$ asymmetry: $\eta = (P'_{yy} - P'_{xx})/(P'_{zz} - P_{iso})$

Interaction Tensors

- In addition to the orientation dependent part of a *spin interaction*, there can be an orientation independent part. Examples are the *indirect spin-spin coupling* and the *chemical shielding*
- The interaction tensor is then the sum of an isotropic part $\mathbf{P}_{iso}$ and an anisotropic part $\mathbf{P}_{aniso}$
- The isotropic part $\mathbf{P}_{iso}$ of the interaction tensor $\mathbf{P} = \mathbf{P}_{iso} + \mathbf{P}_{aniso}$ is given by the trace of the tensor, independent of the coordinate system
- It is observed in liquids where molecules reorient rapidly and randomly
- For symmetric interactions, which are weak compared to the *Zeeman interaction*, the anisotropic part $\mathbf{P}_{aniso}$ is described by the *second Legendre polynomial*
- It is the convention in NMR to measure interaction energies in frequency units according to $E = h\nu = \hbar\omega$
- Usually, several spin interactions act simultaneously and the respective coupling energies are added
- Many NMR methods have been developed with the goal to isolate the effects of one interaction from all the other interactions and to correlate the frequency shifts from different interactions with each other

Energy Levels and Transitions

Energy level diagram

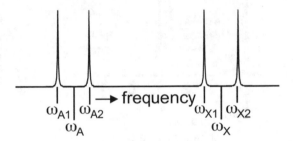

NMR spectrum of
two couped spins with
resonance frequencies
ω_A and ω_X and
orientation-dependent
splittings

Interaction Energies

- The *interaction energies* of coupling nuclei are calculated using *quantum mechanics*
- When the Zeeman interaction dominates, then the interaction energy is essentially proportional to the total spin of the coupled spins
- Depending on the *magnetic quantum numbers* of the interacting partners, a different total spin is obtained and with it a different interaction energy
- The number of nearly equal interaction energies is obtained by combinatorial arguments
- Because the rf excitation employs photons with spin 1, only transitions can be observed, for which the magnetic quantum number m changes by ±1
- These transitions constitute the *transverse magnetization* and are called *single-quantum coherences*
- There, one of the interacting spins changes its orientation in the magnetic field by absorption or stimulated emission of one photon or rf quantum
- For two coupling spins ½, the observable interaction energy differences $\Delta E = \hbar\,\omega$ are proportional to $P_{iso} \pm \delta\,[3\cos^2\beta - 1 - \eta\,\sin^2\beta\,\cos(2\alpha)]/2$ and lead to *orientation-dependent line splittings*
- The absorption or stimulated emission of more than one rf quantum cannot directly be observed in first order
- *Multi-quantum coherences* (0Q, 2Q) can indirectly be observed by *multi-dimensional NMR spectroscopy*

Multiplet Structure in ^{1}H Spectra

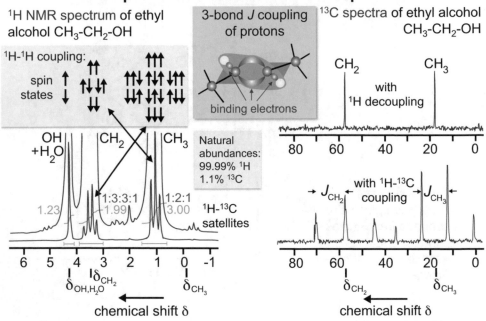

^{1}H NMR spectrum of ethyl alcohol CH_3-CH_2-OH

3-bond J coupling of protons

^{13}C spectra of ethyl alcohol CH_3-CH_2-OH

^{1}H-^{1}H coupling:

spin states

binding electrons

Natural abundances: 99.99% ^{1}H 1.1% ^{13}C

OH +H_2O CH_2 CH_3

1.23   1:3:3:1   1:2:1
        1.99   3.00

^{1}H-^{13}C satellites

6 5 4 3 2 1 0 -1

δ_{OH,H_2O} δ_{CH_2} δ_{CH_3}

chemical shift δ

CH_2 CH_3

with ^{1}H decoupling

80   60   40   20   0

J_{CH_2} with ^{1}H-^{13}C coupling J_{CH_3}

80   60   40   20   0

δ_{CH_2} δ_{CH_3}

chemical shift δ

From B. Blümich, in: R.A. Mayers (ed.) **Enc. Anal. Chem.**, John Wiley, Chichester, 2016, a9458, Fig. 2 with permission

Indirect Spin-Spin Coupling

- Nuclear spins can interact with each other in two ways: one is the *dipole-dipole interaction* through space, the other is the *indirect* or *J coupling*
- The *direct coupling* by the dipole-dipole interaction is described by a traceless tensor **D**, so that the coupling vanishes when averaged by fast motion
- The *indirect spin-spin coupling* is mediated by the electrons of the chemical bonds between the coupling spins
- The nuclear spin interacts with the magnetic field of the orbiting electrons and distorts their paths. This distortion is sensed by the coupling partner spin
- The indirect spin-spin coupling is described by a coupling tensor **J** with trace 3J. The trace is preserved in the fast motion limit
- The coupling leads to *multiplet splittings* of the peaks of the coupled spins
- The sign of the coupling constant J alternates with the number of chemical bonds between the coupling spins
- The multiplet structure bears chemical information and is determined by the different combinations of orientations of the coupling partner spins
- If the total spin of the coupling partner is F, the multiplet has $2F+1$ lines
- Hetero-nuclear J couplings are often exploited for *hetero-nuclear polarization transfer* and chemical *editing of spectra* from molecules in solution
- In contrast to the chemical shift dispersion, the splitting from direct and indirect spin-spin couplings is not removed in the *Hahn echo* maximum

Density Matrix for Two Spins ½

General wave function: $\Psi = a_1(\uparrow\uparrow) + a_2(\downarrow\uparrow) + a_3(\uparrow\downarrow) + a_4(\downarrow\downarrow)$
$= a_1(\alpha\alpha) + a_2(\beta\alpha) + a_3(\alpha\beta) + a_4(\beta\beta)$
$= a_1\psi_1 + a_2\psi_2 + a_3\psi_3 + a_4\psi_4$

Density matrix in the product base:

$$\rho = \begin{matrix} & \alpha\alpha & \beta\alpha & \alpha\beta & \beta\beta \\ \alpha\alpha & \langle a_1a_1^*\rangle & \langle a_1a_2^*\rangle & \langle a_1a_3^*\rangle & \langle a_1a_4^*\rangle \\ \beta\alpha & \langle a_2a_1^*\rangle & \langle a_2a_2^*\rangle & \langle a_2a_3^*\rangle & \langle a_2a_4^*\rangle \\ \alpha\beta & \langle a_3a_1^*\rangle & \langle a_3a_2^*\rangle & \langle a_3a_3^*\rangle & \langle a_3a_4^*\rangle \\ \beta\beta & \langle a_4a_1^*\rangle & \langle a_4a_2^*\rangle & \langle a_4a_3^*\rangle & \langle a_4a_4^*\rangle \end{matrix} \hat{=} \begin{pmatrix} P & 1Q & 1Q & 2Q \\ 1Q & P & 0Q & 1Q \\ 1Q & 0Q & P & 1Q \\ 2Q & 1Q & 1Q & P \end{pmatrix}$$

P: populations, $\omega_{kk} = 0 = |\Delta m|$, leads to longitudinal magnetization
0Q: zero-quantum coherences, $\omega_{kl} \approx 0 = |\Delta m|$, indirectly observable
1Q: single-quantum coherences, $\omega_{kl} \propto 1 = |\Delta m|$, directly observable
2Q: double-quantum coherences, $\omega_{kl} \propto 2 = |\Delta m|$, indirectly observable

Description of NMR with the density matrix $\rho(t) = U(t - t_0) \rho(t_0) U^{-1}(t - t_0)$, the time evolution operator $U(t - t_0) = \exp\{-iH(t-t_0)/\hbar\}$, and the Hamilton operator H

Quantum Mechanics

- The energies E and the *transition frequencies* ω of interacting nuclei are calculated using *quantum mechanics* and including the relation $\Delta E = \hbar \omega$
- Depending on its orientation in the magnetic field, a spin is found in states of different energies E
- Neglecting thermal motion, a spin ½ can be oriented parallel ($\uparrow$, state α) or antiparallel ($\downarrow$, state β) to the polarization field B_0 in thermal equilibrium
- Two coupled spins ½ can assume the four combinations $\uparrow\uparrow, \downarrow\uparrow, \uparrow\downarrow, \downarrow\downarrow$
- States of spins are described in quantum mechanics by *wave functions*
- The wave functions are the *eigenfunctions* of operators similar to the *eigenvectors* of matrices
- The *interaction energies* are the *eigenvalues* of the *Hamilton operator*
- Quantum mechanical operators can be expressed in matrix form
- The *Schrödinger equation* describes the energy balance of a quantum mechanical system by means of the Hamilton operator and wave functions
- The wave function of coupled spins is often expressed as a linear combination of the eigenfunctions of a suitable Hamilton operator
- Accessible by measurement are usually only the ensemble averages of the bilinear products of the complex expansion coefficients
- These averages are written in matrix form and constitute the so-called *density matrix*
- In the eigenbasis of the Hamilton operator the elements of the density matrix are of the general form $A_{mn}\exp\{-(1/T_{mn} + i \Omega_{mn})t\}$

RF Pulse and Free Precession with exp{-i$\Omega_0 t$}

Von Neumann equation of motion of the density matrix:
$d\rho/dt = -i\,[\mathbf{H}, \rho(t)]/\hbar - \Gamma\,[\rho(t) - \rho_0]$

Solution for negligible relaxation ($\Gamma = 0$): $\rho(t) = \rho(t - t_0) = \mathbf{U}(t - t_0)\,\rho(t_0)\,\mathbf{U}^{-1}(t - t_0)$, where $\mathbf{U}(t - t_0) = \exp\{-i\mathbf{H}(t-t_0)/\hbar\}$ is the *time-evolution operator*

Spin-1/2 matrices: $\mathbf{I}_x = 1/2 \begin{pmatrix} 0 & 1 \\ 1 & 0 \end{pmatrix}$, $\mathbf{I}_y = i/2 \begin{pmatrix} 0 & -1 \\ 1 & 0 \end{pmatrix}$, $\mathbf{I}_z = 1/2 \begin{pmatrix} 1 & 0 \\ 0 & -1 \end{pmatrix}$

Boltzmann distribution: $\rho_0 = \exp\{-\mathbf{H}_Z/(k_B T)\}\,/\,\mathrm{Tr}\{\exp[-\mathbf{H}_Z/(k_B T)]\}$
At high temperature: $\rho_0 \propto \mathbf{I}_z$, following $e^{-x} \approx 1 - x$ and $\mathbf{H}_Z = -\hbar\,\mathbf{I}_z\,\gamma\,B_0 = -\hbar\,\mathbf{I}_z\,\Omega_0$

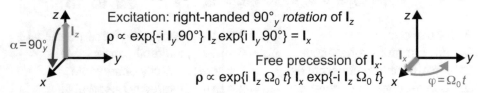

Excitation: right-handed 90°$_y$ *rotation* of $\mathbf{I}_z$
$\alpha = 90°_y$ $\rho \propto \exp\{-i\,\mathbf{I}_y\,90°\}\,\mathbf{I}_z\,\exp\{i\,\mathbf{I}_y\,90°\} = \mathbf{I}_x$

Free precession of $\mathbf{I}_x$:
$\rho \propto \exp\{i\,\mathbf{I}_z\,\Omega_0\,t\}\,\mathbf{I}_x\,\exp\{-i\,\mathbf{I}_z\,\Omega_0\,t\}$ $\varphi = \Omega_0 t$

Expectation value $\langle\mathbf{O}\rangle = \mathrm{Tr}\{\rho\,\mathbf{O}^{\dagger*}\} = \Sigma_{mn}\,\rho_{mn}\,O^*_{nm}$

Transverse magnetization $M_{xy} = M_x + i\,M_y = \hbar\,\gamma\,[\mathrm{Tr}\{\rho\,\mathbf{I}_x^{\dagger*}\} + i\,\mathrm{Tr}\{\rho\,\mathbf{I}_y^{\dagger*}\}] = \hbar\,\gamma\,\rho_{12}$

$M_{xy} \propto \mathrm{Tr}[\exp\{i\,\mathbf{I}_z\Omega_0 t\}\,\mathbf{I}_x\,\exp\{-i\,\mathbf{I}_z\Omega_0 t\}\,\mathbf{I}_x^{\dagger*}] + i\,\mathrm{Tr}[\exp\{i\,\mathbf{I}_z\Omega_0 t\}\,\mathbf{I}_x\,\exp\{-i\,\mathbf{I}_z\Omega_0 t\}\,\mathbf{I}_y^{\dagger*}] = \tfrac{1}{2}\exp\{-i\,\Omega_0 t\}$

Example: Transverse Magnetization

- The time-dependent *Schrödinger equation* transforms into the *density matrix equation of motion* i $d\rho/dt = 1/\hbar\,[\mathbf{H},\rho]$, where $[\mathbf{H},\rho] = \mathbf{H}\,\rho - \rho\,\mathbf{H}$ is the *commutator* of the two operators. It has the solution $\rho(t) = \mathbf{U}(t - t_0)\,\rho(t_0)\,\mathbf{U}^{-1}(t - t_0)$
- At longer evolution times relaxation must be included with the *relaxation super-operator* Γ
- For spins ½ the components of the spin vector operator are proportional to the *Pauli spin matrices*
- In the eigenbase of the Zeeman operator $\mathbf{H}_Z = -\hbar\,\gamma\,\mathbf{I}_z\,B_0$, the *thermodynamic equilibrium density matrix* ρ_0 is given by the *Boltzmann distribution*
- An α_y pulse is given by the evolution operator $\mathbf{R}_y(\alpha) = \exp\{-i\,\mathbf{I}_y\,\alpha\}$, whereby a matrix in the exponent can be evaluated using the expansion $e^x = \Sigma_{n=0}\,x^n/n!$
- A 90°$_y$ *pulse* rotates $\mathbf{I}_z$ into $\mathbf{I}_x$ in the *rotating coordinate frame*
- *Free precession* is a *rotation* of $\mathbf{I}_x$ with frequency Ω around the z axis of the RCF described by the evolution operator $\mathbf{U}(t) = \exp\{-i\,\mathbf{H}/\hbar\,t\} = \exp\{i\,\mathbf{I}_z\,\Omega_0\,t\}$
- The transverse magnetization components M_x and M_y are *ensemble averages* of all spins obtained from the *expectation values* of the $\mathbf{I}_x$ and $\mathbf{I}_y$ spin operators, respectively
- Expectation values of an operator $\mathbf{O}$ are calculated from the product of the transpose (†) and complex conjugate (*) of the operator with the density matrix ρ in terms of the sum of all diagonal elements of the product matrix

Pake Spectrum

$P_{iso} = 0, \quad \eta = 0:$

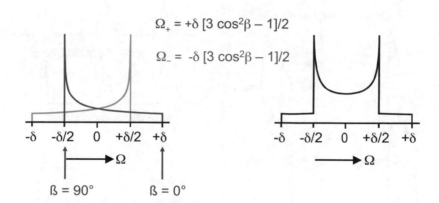

$$\Omega_+ = +\delta\,[3\cos^2\beta - 1]/2$$

$$\Omega_- = -\delta\,[3\cos^2\beta - 1]/2$$

Wideline Spectroscopy

- The *NMR frequency* depends on the orientation of the *interaction tensor* ellipsoid in the laboratory frame
- Consequently, also the separation of the lines or the *multiplet splitting* depends on the *molecular orientation*
- For two dipolar coupled spins ½ and for most deuterons ($I = 1$), $\eta = 0$. In these cases the orientation dependence of the NMR frequency is determined by the *second Legendre polynomial*, $\Omega_\pm = \omega_\pm - \omega_0 = \pm\delta\,(3\cos^2\beta - 1)/2$
- At the *magic angle* $\beta = \arccos(3^{-1/2}) = 54.7°$, $\Omega_\pm = 0$, and only one line is observed at one and the same frequency $\Omega_\pm = 0$ for both transitions
- At other angles, one separate line is observed for each transition with a frequency separation $\Delta\Omega = \omega_+ - \omega_- = \delta\,(3\cos^2\beta - 1)$
- In disordered solids with a range of angles, one obtains a distribution of resonance frequencies, the so-called *wideline spectrum*
- For an *isotropic distribution* it is called a *powder spectrum*
- If $P_{iso} = 0$, $\eta = 0$, and $I = 1$, the powder spectrum is called *Pake spectrum*
- It is observed for pairs of dipolar coupled spins ½ and often also for deuterons
- The Pake spectrum consists of a sum of two wide lines with mirror symmetry, which are centered at the isotropic chemical shift

Wideline-NMR Spectra

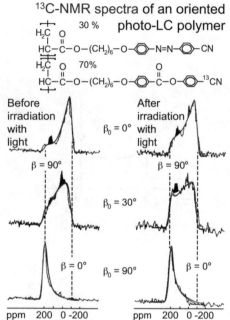

^{13}C-NMR spectra of an oriented photo-LC polymer

Before irradiation with light $\beta_0 = 0°$ After irradiation with light

$\beta = 90°$ $\beta = 90°$

$\beta_0 = 30°$ $\beta = 0°$

$\beta = 0°$ $\beta_0 = 90°$ $\beta = 0°$

ppm 200 0 -200   ppm 200 0 -200

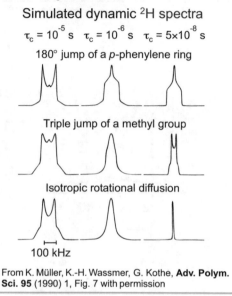

Simulated dynamic ^{2}H spectra

$\tau_c = 10^{-5}$ s $\tau_c = 10^{-6}$ s $\tau_c = 5\times10^{-8}$ s

180° jump of a p-phenylene ring

Triple jump of a methyl group

Isotropic rotational diffusion

100 kHz

From K. Müller, K.-H. Wassmer, G. Kothe, **Adv. Polym. Sci. 95** (1990) 1, Fig. 7 with permission

From A.R. Grimmer, B. Blümich, in: B. Blümich, guest ed., *NMR 30*, **Springer**, Berlin, 1994, p. 3, Fig. 27, with permission

Molecular Order and Dynamics

- The position of a line or the splitting of lines in orientation-dependent spectra can be used to measure *molecular orientations*
- The frequency axis belonging to either of the overlapping wings of a *Pake spectrum* relates to the orientation angle through the *second Legendre polynomial* $P_2(\cos\beta) = \frac{1}{2}(3\cos^2\beta - 1)$
- The angle β is the angle enclosed by the principal axis Z of the *interaction tensor* ellipsoid and the magnetic field B_0
- For *partially oriented solids*, information about the *distribution of orientations* of the interaction tensors is obtained from the lineshape of the spectrum, which also depends on the orientation angle β_0 of the sample in the field B_0
- In case of *molecular motions* with *correlation times* on the time scale of the NMR experiment, the lineshape is altered in a specific way depending on the time scale and geometry of the motion
- *Wideline NMR spectroscopy* is used to analyze the timescale and geometry of *slow molecular motions* in the solid state
- Particularly important is *deuteron wideline NMR spectroscopy* for investigating the *molecular dynamics* of chemical groups labeled site selectively with ^{2}H by exploring the anisotropy of the *nuclear quadrupole interaction*
- Also the anisotropic ^{13}C chemical shift can be explored in the solid state

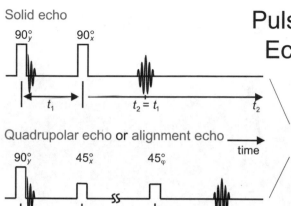

Solid echo

Quadrupolar echo or alignment echo

Pulse Sequences for Echoes from Solid Samples

In the echo maximum the dipole-dipole interaction between two spins ½ is refocused

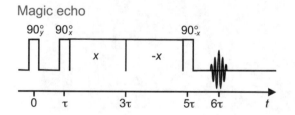

Magic echo

In the echo maximum the dipole-dipole interaction between several spins ½ is refocused

Echoes in Solid Samples

- *Wideline NMR spectra* are favorably acquired by measuring their Fourier transforms in the time domain in terms of *echoes*
- Because the linewidth is broad, the echo signal is narrow
- The time lag between the last rf pulse and the echo serves to overcome the receiver *deadtime* (ca. 5 µs)
- For a system with total spin $I = 1$ (^2H or two dipolar coupled spins ½), maximum echo amplitudes are obtained with the *solid echo* and with the *alignment echo*
- These *solid-state echoes* correspond to the *Hahn echo* and the *stimulated echo* observed for non-interacting spins ½ in liquids
- The refocusing pulses have flip angles with half the values of their liquid-state counter parts, and they are phase shifted by 90° with respect to the first rf pulse
- If more than two spins couple through the dipole-dipole interaction, the *magic echo* leads to maximum echo amplitude
- One example is the dipolar interaction between the three protons of a methyl group
- In the echo maximum, the precession phases of the magnetization components assume their initial values as a result of their interaction during the echo time
- It is said, that the interaction is refocused in the echo maximum

MAS NMR (Magic Angle Spinning)

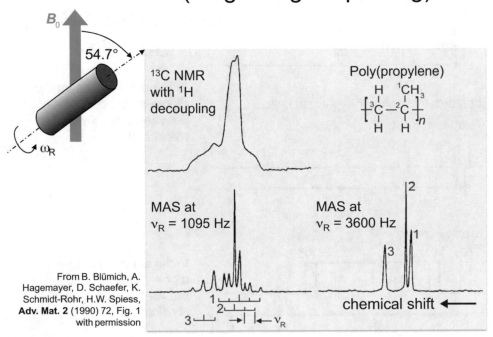

From B. Blümich, A. Hagemayer, D. Schaefer, K. Schmidt-Rohr, H.W. Spiess, **Adv. Mat. 2** (1990) 72, Fig. 1 with permission

Sample Rotation at the Magic Angle

- The *angle dependence of the NMR frequency* resulting from an *anisotropic interaction* is given by $\Omega_{aniso} = \delta [3 \cos^2\beta - 1 - \eta \sin^2\beta \cos(2\alpha)]/2$
- *Sample rotation* in the laboratory at an axis inclined about the angle θ with respect to $\boldsymbol{B}_0$ modulates Ω_{aniso} by $P_2(\cos\theta) = (3 \cos^2\theta - 1)/2$
- The angle dependent part can be eliminated on the time average by rotating the interaction tensors rapidly around the *magic angle* $\theta_m = \arccos\{1/3^{1/2}\} = 54.7°$, where $P_2(\cos\theta_m) = 0$
- 'Rapid' means that the angular *rotation* speed $\omega_R = 2\pi \, \nu_R$ is larger than the principal value δ (in frequency units) of the interaction tensor
- For slower rotation speeds, *spinning sidebands* are observed in the NMR spectrum. These are separated from the isotropic resonance frequency by multiples $n \, \omega_R$ of the spinning speed
- In the limit of vanishing spinning speed, the envelope of the sideband spectrum assumes the shape of the *powder spectrum*
- Sample rotation at the magic angle is called *magic angle spinning (MAS)*
- One of the most important applications of MAS is the measurement of high-resolution ^{13}C-NMR spectra of solid samples
- To this end, the *hetero-nuclear dipole-dipole coupling* between ^{13}C and ^1H is typically eliminated by *dipolar decoupling* (DD) through rf irradiation of ^1H while the anisotropy of the ^{13}C chemical shift is removed by MAS

CPMAS

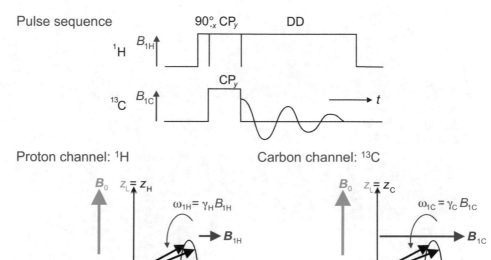

Cross-Polarization, MAS, and Hetero-Nuclear Dipolar Decoupling

- To measure *high-resolution solid-state NMR spectra* of rare nuclei such as ^{13}C and ^{29}Si the *anisotropy of the chemical shift* needs to be eliminated by *MAS* and the *hetero-nuclear dipole-dipole coupling* to ^{1}H by *dipolar decoupling*
- To decouple rare nuclei like ^{13}C from abundant nuclei like ^{1}H, the abundant nuclei are irradiated with particular pulse sequences or simply a strong B_1 field while observing the rare nuclei
- Due to the fact that the natural abundance of ^{13}C is only 1%, and $\gamma_{13C} \approx \gamma_{1H}/4$, the thermal equilibrium polarization of ^{13}C is much lower than that of ^{1}H
- Furthermore, the T_1 relaxation time of ^{13}C is usually longer than that of ^{1}H
- Both disadvantages can be alleviated by transfer of magnetization from ^{1}H to ^{13}C with a method called *cross polarization (CP)*
- To this end, one simultaneously irradiates resonant B_1 fields to ^{1}H and ^{13}C with amplitudes critically chosen, so that the ^{1}H magnetization as well as the ^{13}C magnetization rotate around their individual B_1 fields with the same frequency
- This adjustment fulfills the *Hartmann-Hahn condition*: $\gamma_H B_{1H} = \gamma_C B_{1C}$
- Then, parallel to $\boldsymbol{B}_0$, the ^{1}H and ^{13}C spins oscillate with the same frequency in the laboratory as well as in the rotating coordinate frames
- By this resonance effect, transverse ^{1}H magnetization can be converted directly into transverse ^{13}C magnetization

Homo-Nuclear Dipolar Decoupling

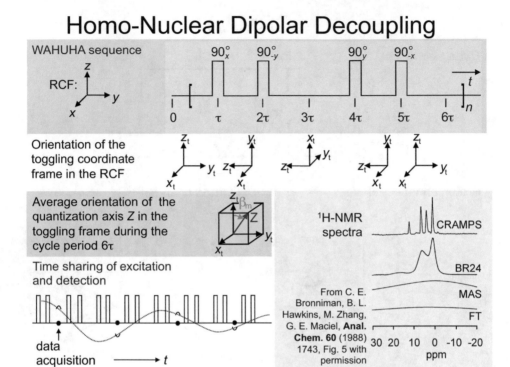

Solid-State Multi-Pulse NMR

- *Energy levels* and *NMR frequencies* are calculated with the *Hamilton operator* and the *density matrix* following the rules of *quantum mechanics*
- The expression for the Hamilton operator consists of a space- and a spin-dependent part
- The space-dependent part describes the *interaction anisotropy*
- The spin-dependent part determines the constitution of the energy level diagram including the observable transitions
- The *orientation dependence of the NMR frequency* can be eliminated by manipulation of the space-dependent part using *MAS*, but also by manipulation of the spin-dependent part using *multi-pulse NMR*
- The simplest *multi-pulse sequence* for elimination of the *dipole-dipole interaction* is the *WAHUHA sequence* by Waugh, Huber, and Haeberlen
- It consists of four 90° pulses and is cyclically repeated
- In each cycle, one data point is acquired stroboscopically at time 3τ or 6τ
- The pulse cycle is designed in such a way, that the quantization axis of the Hamilton operator is aligned along the space diagonal on the time average
- This axis encloses the *magic angle* with the z axis of the rotating frame RCF
- Homo-nuclear dipolar decoupling by MAS is improved by combining multi-pulse NMR (e. g. BR24) and MAS yielding *CRAMPS*: combined rotation and multi-pulse spectroscopy

Double-Quantum NMR

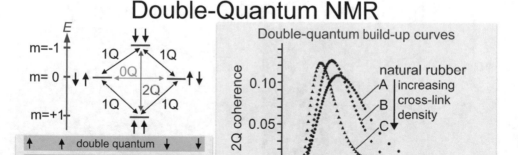

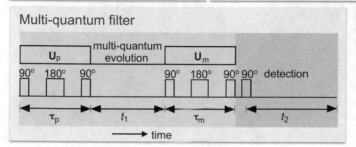

From M. Schneider, D.E. Demco, B. Blümich, **J. Magn. Reson. 140** (1999) 432, Fig. 3b with permission

Multi-Quantum NMR

- *Multi-quantum coherences* are superposition states with $|\Delta m| \neq 1$, where, for example, two or more interacting spins ½ flip simultaneously
- Multi-quantum coherences can be detected only indirectly via a modulation of directly detectable *single-quantum coherences* (*transverse magnetization*)
- They are generated from equilibrium magnetization with at least two rf pulses
- To suppress the chemical-shift evolution between the pulses, a 180° pulse is centered in the *preparation period* to form a *Hahn echo* at its end
- The resultant sequence of three rf pulses and two precession intervals is called the *preparation propagator* U_p of duration τ_p
- In a subsequent *multi-quantum evolution period* t_1 the multi-quantum coherences precess and relax similar to transverse magnetization
- For detection they are converted into directly observable single-quantum coherences or first into *longitudinal magnetization* by the *mixing propagator* U_m
- For longitudinal magnetization U_m is a time reverse copy of the preparation propagator U_p
- The *multi-quantum coherence order* can be selected by the pulse phases in combination with suitable phase cycling schemes for signal accumulation
- The *multi-quantum build-up curves* (signal amplitude versus $\tau_p = \tau_m$) are steep in the initial part for strong *dipole-dipole interactions*
- Their decay is governed by relaxation
- Multi-quantum pulse sequences can serve as *dipolar filters* to select magnetization from rigid domains of dynamically heterogeneous solids

Principle of 2D NMR Spectroscopy

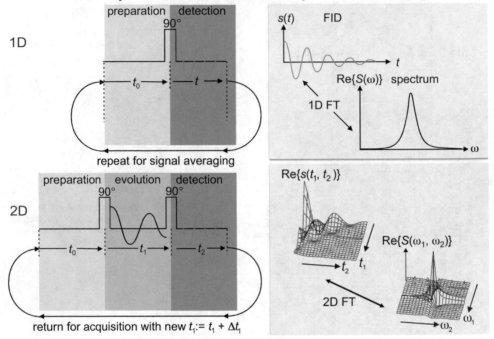

Introduction to Multi-Dimensional NMR Spectroscopy

- *Multi-dimensional NMR spectroscopy* concerns the generation of NMR spectra with more than one frequency axis
- Multi-dimensional Fourier spectra are generated by measuring FIDs with amplitudes or phases that are modulated by coherences of the density matrix
- For example, several pulses can be applied and the delays between them varied before the signal is acquired after the last pulse
- In *2D NMR spectroscopy*, FIDs are acquired following at least two rf pulses with increasing *evolution time* between the pulses
- The FIDs are stored in the rows of a data matrix
- A *2D spectrum* is obtained by *2D Fourier transformation* of the data matrix
- A 2D FT consists of 1D FTs for all rows and all columns of the matrix
- Straight forward 2D FT leads to 2D peaks with a phase twist, which cannot be phase corrected
- Purely absorptive 2D peaks are obtained by suitable phase cycling and data manipulation
- Depending on the pulse sequence, correlations between different peaks in 1D spectra are revealed (*correlation NMR*), or crowded 1D spectra are thinned out by spreading them out in two or more dimensions (*separation NMR*)

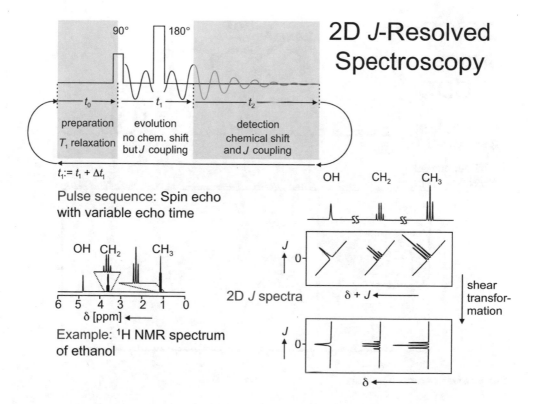

2D *J*-Resolved Spectroscopy

Pulse sequence: Spin echo with variable echo time

2D *J* spectra

Example: ¹H NMR spectrum of ethanol

shear transfor-mation

A Simple Example of 2D NMR Spectroscopy

- The simplest example for 2D spectroscopy is *J-resolved spectroscopy*
- In 2D *J*-resolved NMR spectra, the *J multiplets* appear in one dimension and the *chemical shift* δ in the other
- The basic pulse sequence is the *spin echo* sequence
- In the echo maximum, the evolution of the transverse magnetization due to chemical shift and magnetic field inhomogeneity is refocused, while the evolution before and after is governed by chemical shift and *J* coupling
- Choosing the echo time as the evolution time t_1, and acquiring the decay of the echo during the detection time t_2 leads to a 2D data matrix. After *2D Fourier transformation*, the spin multiplets are centered in the second dimension at $J = 0$ and spread along an axis oriented at 45° in the plane
- A shear transformation aligns the *J* multiplets along one axis, so that *J* coupling and chemical shift are separated in both dimensions of the spectrum
- *2D J-resolved spectroscopy* is an example of *2D separation NMR*. The 1D spectrum is spread into a second dimension, but no new peaks are observed
- Due to a *phase twist*, a 1D projection of the 2D spectrum onto the chemical-shift axis results in zero signal amplitude
- To obtain a ¹H decoupled 1D spectrum with resonances affected by chemical shift only and not by the *J*-coupling (pure shift spectrum), a projection is calculated across the *J*-dimension from the magnitude of the 2D spectrum

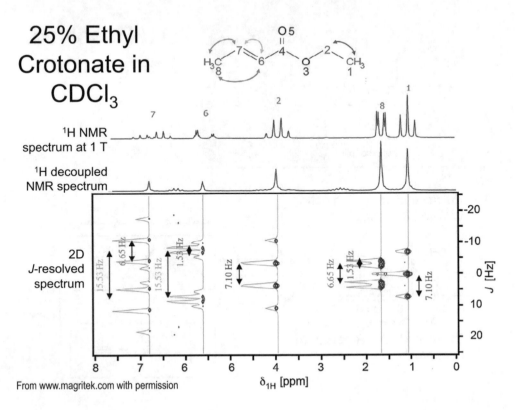

From www.magritek.com with permission

Example: 2D *J*-Resolved Spectroscopy of Ethyl Crotonate

- Ethyl crotonate contains 5 chemical groups with protons giving rise to 5 multiplets in the ¹H NMR spectrum
- The mutiplets are centered at the chemical shifts in the conventional 1D spectrum
- In the *J-resolved 2D spectrum* the multiplets are rotated into the second dimension so that the multiplet splitting is no longer observed along the chemical shift dimension
- Whenever the coupling is of the same order of magnitude as the chemical shift difference between the coupling partners, the strong coupling regime is met, and combination lines arise outside the *J*-coupling range shown in the 2D spectrum
- A projection of the magnitude of the 2D spectrum along the *J*-coupling dimensions yields the *homo-nuclear decoupled 1D spectrum*
- Due to the magnitude mode the lines in this 1D spectrum are broader than in the phase-sensitive *J*-coupled 1D spectrum

2D COSY and Multi-Quantum NMR Spectroscopy

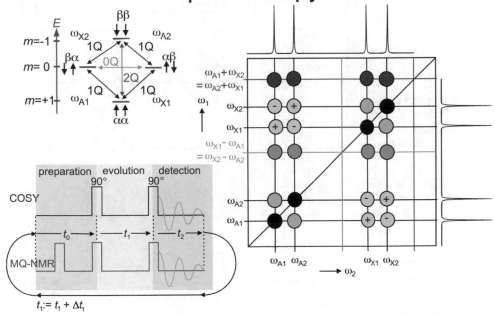

Multi-Dimensional Correlation NMR Spectroscopy

- The simplest pulse sequence for uncovering correlations between lines in 1D spectra is the double-pulse experiment with two 90° pulses
- It generates correlation spectra (*COSY: correlation spectroscopy*), in which coupled lines in 1D spectra are identified by cross-peaks
- Depending on the *connectivity* in the energy level diagram, whether progressive ($\Delta m = \pm 2$) or regressive ($\Delta m = 0$), the phase of the cross-peaks is negative and positive, respectively
- When preparing the initial state of the spin system with two or more pulses, *multi-quantum coherences* are excited in the evolution period, which modulate the single-quantum coherences recorded in the detection period
- The primitive *multi-quantum pulse sequence* is obtained from the basic two-pulse sequence by starting from thermodynamic equilibrium and preparing the spin system for the evolution period with two 90° pulses instead of one
- There is a wide variety of pulse sequences for generating different *multi-dimensional NMR spectra* sensitive to homo- and hetero-nuclear couplings
- Multi-dimensional NMR spectroscopy experiments are best analyzed with the quantum mechanical *density matrix* formalism

Homo-Nuclear 2D Correlation NMR Spectroscopy

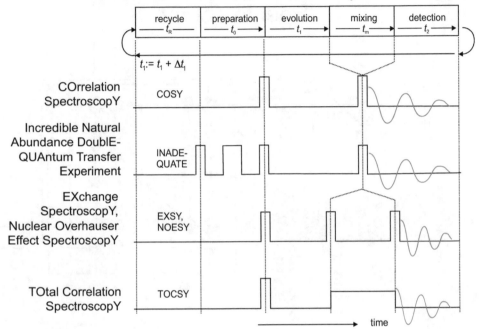

2D Correlation NMR Spectroscopy

- All sequences of multi-dimensional NMR with pulsed excitation follow a general scheme of building blocks lasting for certain durations or time periods
- For 2D NMR the following time periods succeed each other: *recycle period* or *recovery period*, *preparation period*, *evolution period*, *mixing period*, and *detection period*
- Depending on the particular pulse sequence, certain periods can be omitted
- The recycle period t_R serves to establish a defined initial magnetization. Typically this is the *thermodynamic equilibrium polarization* and $t_R \approx 5\ T_1$
- During the preparation period the nuclear polarization can be converted to particular coherences in the density matrix, e.g. to *double-quantum coherences* in the *INADEQUATE* experiment
- During the evolution period the first dimension of the 2D spectrum is encoded. This is done most frequently by incrementing t_1 from scan to scan
- The mixing period serves to mix coherences, which belong to the same spin system and are encoded during t_1 either coherently by particular blocks of rf pulses or incoherently by temperature-activated processes
- During the detection period the transverse magnetization decay is sampled as a function of the detection time t_2, which encodes the second dimension
- Popular homo-nuclear 2D experiments are the *COSY* experiment, the *multi-quantum experiment* known as *INADEQUATE* in ^{13}C NMR, the *EXSY* experiment and the *NOESY* experiment

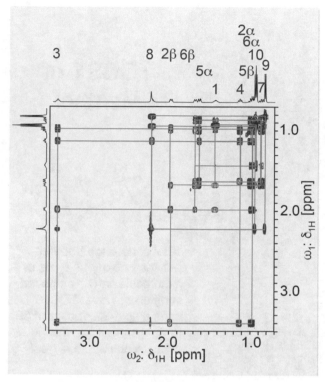

COSY of Menthol

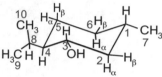

COSY spectrum at 7 T for identification of coupled spins

2D Correlation Spectroscopy: COSY

- The 2D *COSY* (COrrelation SpectroscopY) experiment is the 2D NMR experiment originally proposed by J. Jeener in 1971
- It is generated by applying two 90° rf pulses instead of just one which are separated by a variable *evolution time* t_1
- The first pulse and the subsequent evolution time t_1 prepare the spin system in a non-equilibrium state, which is probed by recording the *impulse response* (*FID*) after the second pulse during the detection time t_2 as a function t_1
- The Fourier transform of the time-domain data set $s(t_1, t_2)$ is a *2D spectrum*
- Both its axes bear the same information as the homo-nuclear 1D spectrum so that the 1D spectrum appears along the diagonal
- Off-diagonal peaks identify lines which are coupled and belong to the same spin system
- In liquid-state NMR the most significant coupling is the *indirect* or *J coupling*
- To resolve the *multiplet splitting*, the evolution and detection times have to be at least as long as the inverse coupling constant
- Often, magnitude spectra are displayed and the line-shapes are artificially adjusted from star-shaped to circular-shaped 2D peaks by changing the envelope of the time-domain signal prior to *2D Fourier transformation*

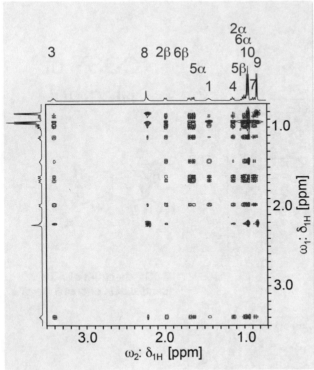

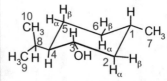

TOCSY of Menthol

Total coherence transfer between nearly all spins of a molecule can be achieved in different ways. TOCSY uses a spin-lock pulse of 50 to 75 ms after the mixing pulse to achieve this goal

Total Correlation Spectroscopy: TOCSY

- In the regular *COSY* spectrum, coupled spins like A and B as well as B and C are identified by cross-peaks
- If the coupling between A and C is too weak, no cross-peak is observed although all three spins A, B, and C belong to the same network of spins
- To identify different networks of spins in crowded spectra, it is helpful to generate a COSY-type spectrum which shows cross-peaks between all spins of a network by relaying the magnetization of spin A to spin C and *vice versa* via spin B
- Experiments of this type are called *TOCSY* (TOtal Coherence transfer SpectroscopY) experiments
- In the simple form of TOCSY the 90° mixing pulse of the COSY experiment is replaced by a spin-lock period of 50 to 75 ms duration, in which all coupled transitions are mixed, because chemical shift is not resolved in the B_1 field
- The TOCSY experiment is a standard tool in the structural analysis of biological macromolecules by multi-dimensional high-resolution NMR spectroscopy
- The TOCSY spectrum of menthol shows many more cross-peaks than the corresponding COSY spectrum

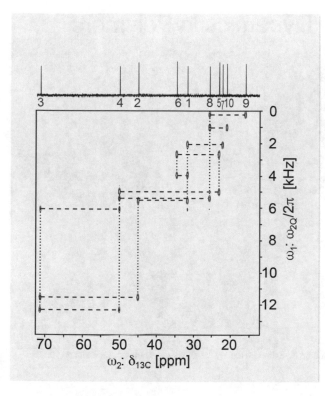

INADEQUATE of Menthol

Direct carbon-carbon
connectivity through
double-quantum NMR
spectroscopy of ^{13}C

2D Double-Quantum NMR Spectroscopy

- A *COSY* spectrum may not provide all the information needed to unambiguously assign the resonance lines to the chemical structure of a molecule
- ^{13}C NMR provides better *chemical shift resolution* than ^{1}H NMR but is less sensitive due to the low natural abundance of ^{13}C (1%) and its gyromagnetic ratio being lower than that of ^{1}H
- *J*-coupled ^{13}C spin pairs arise with a probability of 0.0001 (1% of 1%)
- Nevertheless, they can be detected at high field, and *double-quantum coherences* can be generated in such spin pairs
- *INADEQUATE* is the 2D version of the double-quantum ^{13}C NMR experiment
- The indirectly detected ω_1 dimension is a double-quantum frequency corresponding to the sum frequency of the coupling spins; the directly detected ω_2 dimension corresponds to the chemical shift in the 1D spectrum
- Along ω_2, pairs of doublets are observed, centered at the chemical shifts of the directly bonded ^{13}C spins in the carbon backbone of the molecule
- The pairs of doublets at frequencies ω_1 of a particular carbon lead to its different carbon neighbors in the ω_2 dimension
- With 2D INADEQUATE NMR spectroscopy the complete carbon skeleton of a molecule can be traced due to the high sensitivity at high magnetic field and the large dispersion of the ^{13}C chemical shift

Segmental Dynamics in Polymers

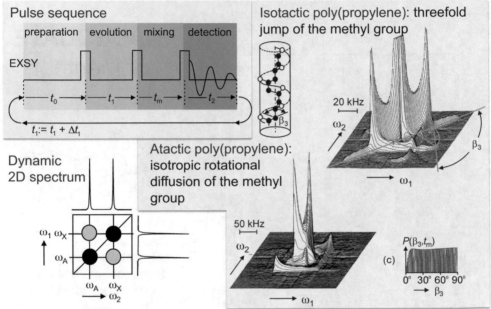

From B. Blümich, A. Hagemeyer, D. Schaefer, K. Schmidt-Rohr, H. W. Spiess, **Advanced Materials 2** (1990) 72, Figs. 3, 4, 5 with permission

Dynamic 2D NMR Spectroscopy

- *Dynamic multi-dimensional NMR spectroscopy* generates spectra, which approximate combined probability densities for a spin packet with an initial NMR frequency to have another final NMR frequency after a *mixing time* t_m
- Initial and final NMR frequencies are labeled in the evolution period t_1 and the detection period t_2, respectively. In the *slow-motion limit*, these are so short compared to t_m, that no appreciable motion arises during these periods
- Dynamic processes or motions relevant to NMR spectroscopy are *rotations* of chemical groups in liquids, which are associated with a change in NMR frequency, reorientations of molecules in solids with an anisotropic chemical shift, and cross-relaxation corresponding to the *nuclear Overhauser effect* (*NOE*, see below)
- Exchange NMR spectroscopy (*EXSY*) leads to cross-peaks at the cross-coordinates of initial and final frequencies
- In powders and partially oriented systems, *wideline exchange NMR spectra* are observed: the off-diagonal signals provide detailed information about the geometry and timescale of the molecular motion
- By modeling the solid-state exchange NMR spectrum, one obtains the *distribution* $P(\beta_3, t_m)$ *of reorientation angles* β_3 which are accessed by the particular type of motion during the mixing time t_m

Exchange, Coalescence, and Motional Narrowing

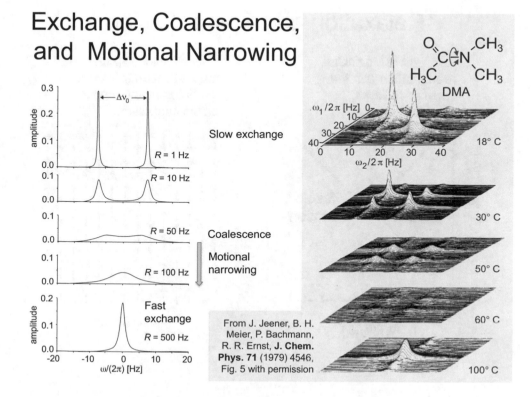

Slow exchange

$R = 1$ Hz

$R = 10$ Hz

$R = 50$ Hz   Coalescence

$R = 100$ Hz   Motional narrowing

Fast exchange

$R = 500$ Hz

DMA

18° C

30° C

50° C

60° C

100° C

From J. Jeener, B. H. Meier, P. Bachmann, R. R. Ernst, **J. Chem. Phys. 71** (1979) 4546, Fig. 5 with permission

Exchange NMR Spectroscopy in Liquids

- For liquids 2D *exchange cross peaks* can arise from hindered rotations around chemical bonds
- The classical examples are *N,N*-dimethylformamide (DMF) and dimethyl-acetamide (DMA), where the chemical groups rotate around the C-N bond, so that the *cis* and *trans* methyl groups exchange their positions
- The life times $\tau_c = k_{c \to t}^{-1}$ and $\tau_t = k_{t \to c}^{-1}$ depend on temperature, where $R = k_{c \to t} + k_{t \to c}$ is the rate of the exchange process
- Depending on the ratio R of exchange rate and frequency separation $\Delta\nu_0 = (\delta_c - \delta_t)\,\omega_0/(2\pi)$ of the resonances, one or two lines are observed in the spectrum (note: δ = 2.74 and 2.91 ppm for dimethylacetamide)
- In the fast exchange limit at $R > 100$ Hz, one line is observed
- Near $R = 50$ Hz, the lines coalesce and become small and broad
- In the slow exchange limit for $R < 1$ Hz, two lines are observed in the 1D spectrum, and for mixing times $t_m > 1/R$, cross-peaks are observed in the *2D exchange spectrum*
- The exchange cross peaks exhibit the same phase as the auto peaks on the diagonal

Relaxation and Spin Diffusion

Relaxation: Molecular
reorientation with the
correlation time τ_c

Spin diffusion: Spatial
magnetization transport
mediated by the dipole-
dipole interaction

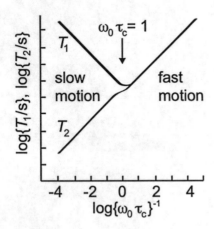

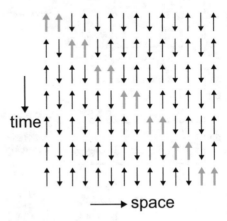

Dynamic Processes

- The decay of transverse magnetization and the build-up of longitudinal magnetization are determined by the *relaxation times* T_2 and T_1, respectively
- In homogeneous samples the build-up of longitudinal magnetization proceeds in an exponential fashion for liquids and solids
- *Transverse relaxation* is often Gaussian in solids and exponential in liquids in the limit of fast molecular motion
- The dominating relaxation mechanism is the time-dependent *dipole-dipole interaction* between a nuclear spin and other magnetic dipoles such as paramagnetic centers (free electrons) on neighboring chemical groups or molecules
- A time-dependent modulation of a spin coupling is achieved primarily by *rotational motion* but also by *translational motion*
- In the *fast motion limit*, the motions at frequencies ω_0 and $2\omega_0$ determine the nuclear T_1 relaxation and those at frequencies 0, ω_0 and $2\omega_0$ the T_2 relaxation
- This is why T_1 and T_2 differ for slow motions
- In addition to the dipole-dipole coupling to electrons, the dipole-dipole coupling to other nuclei, the *anisotropy of the chemical shift*, and in gases, the *spin rotation interaction* are active nuclear magnetic relaxation mechanisms
- *Longitudinal magnetization* moves towards spatial equilibrium by *spin diffusion*, which denotes the migration of magnetization by energy conserving flip-flop transitions of coupled spins

Relaxation Paths in a Two-Spin IS System

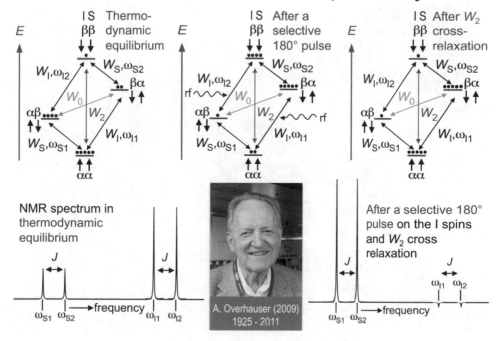

The Nuclear Overhauser Effect: NOE

- In 1955 A.W. Overhauser suggested to saturate the resonance of unpaired electrons to enhance the NMR signal of spins coupled to the electrons
- If the T_1 relaxation is governed by the dipole-dipole interaction, and if there is appreciable *cross-relaxation* between coupling spins S and I, the *Overhauser effect* can be observed
- It can be used to amplify the signal of a weakly polarized species S by cross-relaxation from a highly polarized species I following a perturbation of the thermodynamic equilibrium magnetization of the I spins
- A perturbation of the S spins is achieved, e.g., by selective *inversion* of the I-spin polarization with a 180° pulse or in a systematic fashion with 2D NMR
- The signal amplitudes are given by the population differences of the energy levels defining the transition frequencies
- If the *cross-relaxation rate* W_2 is strong, the signal amplitude of the S spins changes by up to $S_z/S_0 = 1 + \eta$, where η is the *enhancement factor*
- η depends on the relaxation rates W_i, $\eta = \gamma_I(W_2-W_0)/[\gamma_S(2W_S+W_2+ W_0)]$, whereby $W_2-W_0 \propto \tau_c (r_{IS})^{-6}$. It is proportional to the correlation time τ_c of molecular motion and to $(r_{IS})^{-6}$, where r_{IS} is the distance between the cross-relaxing spins
- The maximum enhancement is given by $\eta = \gamma_I/(2\gamma_S)$
- The *NOE* is also used for determining proximity of spins in isotropic fluids
- For partially oriented molecules, such as molecules residing in the anisotropic pores of a stretched gel, the *residual dipole-dipole coupling* can also serve to characterize distances between spins

2D NOE Spectroscopy (NOESY)

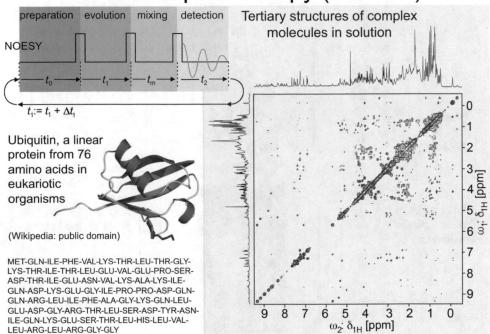

Tertiary structures of complex molecules in solution

Ubiquitin, a linear protein from 76 amino acids in eukariotic organisms

(Wikipedia: public domain)

MET-GLN-ILE-PHE-VAL-LYS-THR-LEU-THR-GLY-
LYS-THR-ILE-THR-LEU-GLU-VAL-GLU-PRO-SER-
ASP-THR-ILE-GLU-ASN-VAL-LYS-ALA-LYS-ILE-
GLN-ASP-LYS-GLU-GLY-ILE-PRO-PRO-ASP-GLN-
GLN-ARG-LEU-ILE-PHE-ALA-GLY-LYS-GLN-LEU-
GLU-ASP-GLY-ARG-THR-LEU-SER-ASP-TYR-ASN-
ILE-GLN-LYS-GLU-SER-THR-LEU-HIS-LEU-VAL-
LEU-ARG-LEU-ARG-GLY-GLY

Through-Space Distance Information

- The *nuclear Overhauser effect* is often explored in 2D NMR spectroscopy
- The pulse sequence equals that of the *EXSY* experiment but is referred to as *NOESY* for Nuclear Overhauser Effect SpectroscopY
- Compared to EXSY, NOESY uses longer *mixing periods* for *cross-relaxation* and elimination of transverse magnetization components by T_2 relaxation
- The population differences are prepared by the first two pulses and the *evolution time t_1*. They are modified in the *mixing time t_m* and detected via the signal amplitudes of the *FID* recorded during the *detection time t_2*
- The initial perturbation of the populations is systematically varied by incrementing the evolution time through a range of values
- A 2D FT leads to a 2D spectrum with cross peaks due to cross-relaxation
- The cross-peaks provide distance constraints to refine the *tertiary structure* of large molecules in solution, because cross-relaxation is dominated by the through-space *dipole-dipole interaction* of spins 0.18 to 0.5 nm apart
- To interpret NOE spectra, all resonance lines need to have been assigned to the *secondary structure* of the molecule
- This assignment is accomplished with a variety of different homo- and hetero-nuclear *multi-dimensional NMR spectra* typically involving ^{1}H, ^{13}C, and ^{15}N, often from molecules prepared with selective isotope labels
- For better *sensitivity*, experiments with *inverse detection*, i.e. detection of ^{1}H instead of the hetero-nucleus are preferred

Illustration for the *J*-Coupled IS System

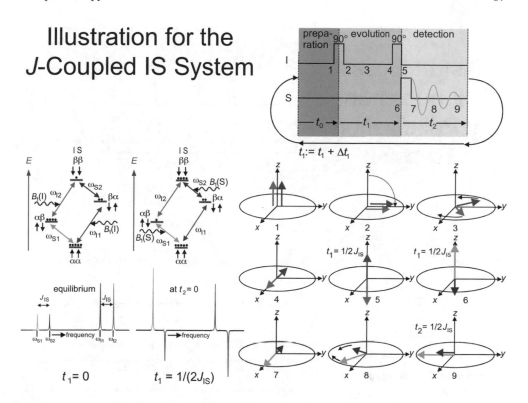

Coherent Hetero-Nuclear Polarization Transfer via Populations of Energy Levels

- Hetero-nuclear experiments explore the *polarization transfer* (*transfer of long-itudinal magnetization*) between different spins I and S, such as ^1H and ^{13}C
- *Coherent polarization transfer* makes use of spin couplings; *incoherent polarization transfer* makes use of T_1 relaxation, e.g. the NOE effect
- In liquids the *hetero-nuclear indirect coupling* is used for coherent transfer
- It makes use of either common energy-levels or multi-quantum coherences common to spins I and S
- To affect the populations of energy levels, transverse I-spin magnetization is generated, and the doublet components are allowed to precess for a time t_1 = $1/(2J_{IS})$ to align in opposite directions in the transverse plane
- These anti-phase components are converted to longitudinal magnetization by a 90° pulse resulting in a redistribution of populations of the energy levels
- The new distribution of populations shows enhanced *population differences* but with opposing signs
- This distribution is interrogated by a 90° pulse applied to the S spins
- The resultant transverse anti-phase S-spin magnetization forms an *echo* after time t_2 = $1/(2J_{IS})$. The Fourier transform of the echo decay is an in-phase doublet
- Hetero-nuclear 2D spectroscopy executes this experiment in a systematic way

DEPT:
Distortionless
Enhancement
by Polarization
Transfer

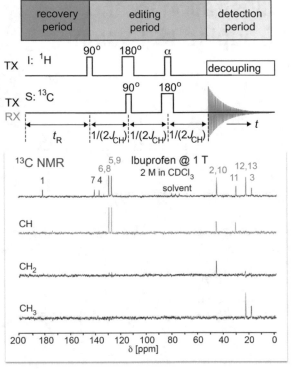

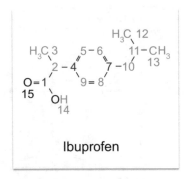

Ibuprofen

From www.magritek.com with permission

Polarization Transfer and Spectral Editing
via Multi-Quantum Coherences

- The transfer of nuclear polarization between coupled spins via *multi-quantum coherences* depends on the multiplet order and the rf-pulse flip angle
- The *DEPT* sequence (*Distortionless Enhancement by Polarization Transfer*) makes use of this to enhance the ^{13}C signal (S-spins) in a way that depends on the number of coupled protons (I-spins)
- The sequence starts with a $90°_x$ pulse to create transverse ^1H magnetization, which evolves for a duration $1/(2J_{CH})$
- The following $180°_x$ pulse inverts the resulting magnetization. Simultaneously multi-quantum coherences are created with a $90°_x$ pulse on the ^{13}C spins
- After another delay $1/(2J_{CH})$ the ^1H chemical shift is refocused
- At that time the ^{13}C magnetization is inverted with a $180°_x$ pulse while the ^1H coherences are rotated with an α-pulse to generate anti-phase magnetization
- After a further delay $1/(2J_{CH})$ the anti-phase magnetization merges to in-phase transverse magnetization and can be decoupled from the ^{13}C spins
- The intensity of the observed ^{13}C signals depends on the flip angle α
- $\alpha = 135°$ creates positive signals for CH and CH_3 and negative signals for CH_2. $\alpha = 90°$ creates signal only for CH and $\alpha = 45°$ from all protonated carbons
- All three spectra properly added and subtracted produce three separate *sub-spectra*, one for each type of group CH, CH_2, and CH_3

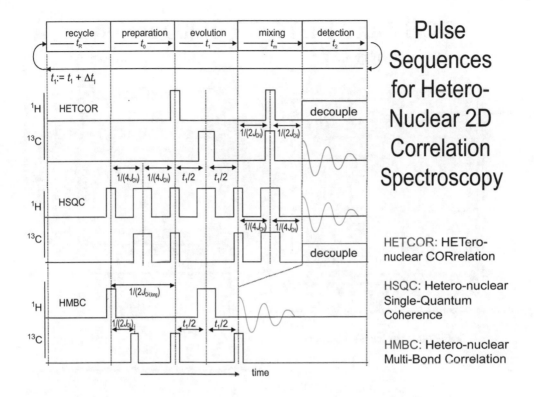

Pulse Sequences for Hetero-Nuclear 2D Correlation Spectroscopy

HETCOR: HETero-nuclear CORrelation

HSQC: Hetero-nuclear Single-Quantum Coherence

HMBC: Hetero-nuclear Multi-Bond Correlation

Hetero-Nuclear 2D Correlation NMR Spectroscopy

- To interpret the lines in a ^{1}H NMR spectrum it is helpful to make use of the large *chemical shift dispersion* of the ^{13}C resonances
- Given enough sample, the lines in a ^{13}C spectrum may be assigned with the help of the *2D INADEQUATE* spectrum via *double-quantum coherences*
- ^{1}H and ^{13}C NMR spectra can be correlated with hetero-nuclear variants of the *COSY* experiment
- They make use of the *hetero-nuclear indirect coupling* between ^{1}H and ^{13}C
- Only ^{1}H nuclei coupled to ^{13}C nuclei bear the connectivity information
- With ^{13}C in natural abundance, 99% of the ^{1}H signal is from protons attached to ^{12}C
- This part of the signal is discarded whenever ^{13}C is detected
- But due to higher resonance frequency, ^{1}H is detected with higher sensitivity. In case of ^{1}H detection the signal from ^{1}H bound to ^{12}C must be eliminated, for example, by phase cycling
- The *HETCOR* experiment is the straight forward extension of the COSY experiment with direct ^{13}C detection
- The *HSQC* and the *HMBC* experiments are hetero-nuclear 2D NMR experiments with direct ^{1}H detection also known as *inverse detection*

COSY and HETCOR of Ibuprofen

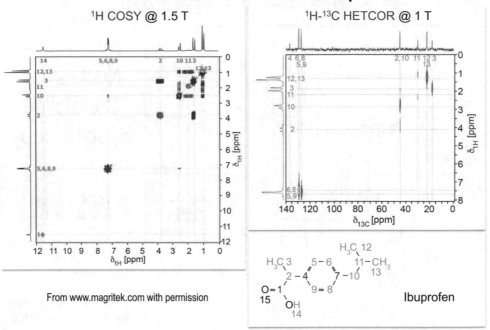

From www.magritek.com with permission

2D Hetero-Nuclear Correlation NMR with ^{13}C Detection

● In the *HETCOR* experiment, transverse ^{1}H magnetization is excited and evolves during the evolution time

● The coupled ^{13}C spins are decoupled from the ^{1}H spins at the end of the evolution time t_1 by generation of a ^{13}C echo

● Following further precession for a time $1(2J_{CH})$ the ^{1}H-^{13}C spin pairs produce anti-phase ^{1}H magnetization, which can be transferred to anti-phase ^{13}C magnetization by *coherent polarization transfer* via populations of energy levels with 90° pulses in the ^{1}H and the ^{13}C channels

● Following another precession period of duration $1(2J_{CH})$ the anti-phase ^{13}C magnetization is refocused and detected as the ^{13}C FID during t_2 while the protons are decoupled with rf irradiation in the ^{1}H channel

● During the de- and refocusing times $1/(2J_{CH})$ appreciable signal may be lost by T_2 relaxation. In practice the delays are optimized for maximum transfer efficiency and minimum signal loss

● Quaternary carbons cannot be observed in HETCOR spectra with indirect detection

HSQC and HMBC of Ibuprofen at 1.5 T

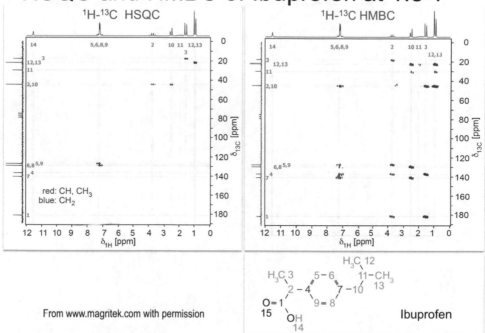

From www.magritek.com with permission

Ibuprofen

Sensitivity Enhancement by ^{1}H Detection

- The *sensitivity* of the *hetero-nuclear 2D NMR* experiments improves and the measurement time considerably shortens if ^{1}H is detected instead of ^{13}C
- The strong signal from ^{1}H at ^{12}C is eliminated in favor of that from ^{1}H at ^{13}C by addition of signals acquired with different phases of the rf pulses
- In the *HSQC* experiment *coherent polarization transfer* of anti-phase ^{1}H magnetization to ^{13}C boosts the sensitivity in a preparation period of duration $1/(2J_{CH})$, where J_{CH} is the one-bond hetero-nuclear coupling
- At the end of t_1 the effect of the J_{CH} coupling on the evolution of the ^{13}C magnetization is eliminated by generating an echo of the ^{1}H magnetization
- During the subsequent mixing time, the ^{13}C magnetization is converted back to ^{1}H magnetization in a second coherent polarization transfer step
- At the beginning of the detection time $t_2 = 0$ the ^{1}H magnetization components are in phase and are detected with ^{13}C decoupling during t_2
- The *HMBC* experiment generates *multi-quantum coherences* first through the strong single-bond J_{CH} coupling and then through the weaker multi-bond $J_{CH,long}$ coupling with two 90° pulses on ^{13}C during the preparation period
- At the end of t_1, the ^{13}C magnetization is converted to anti-phase ^{1}H magnetization and detected without refocusing to reduce signal loss from relaxation
- Cross-peaks from one-bond J_{CH} couplings are eliminated by *phase cycling*
- Cross-peaks arise from long-range couplings $J_{CH,long}$ but they are split due to the lack of ^{13}C decoupling during the detection time t_2

4. Imaging and Transport

The precession phase
Scanning of k space
Slice and volume selection
Spin-echo imaging
Gradient-echo imaging
Spectroscopic imaging
Fast imaging
Velocity fields
Velocity distributions
Exchange NMR
Projections and cross-sections

© Springer Nature Switzerland AG 2019
B. Blümich, *Essential NMR*,
https://doi.org/10.1007/978-3-030-10704-8_4

The Gradient-Field with a Constant Field-Gradient

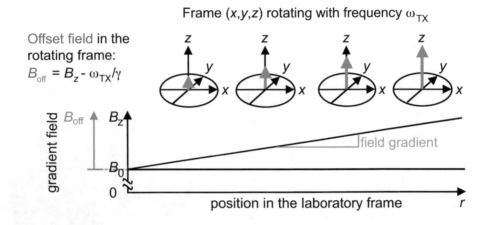

Frame (x,y,z) rotating with frequency ω_{TX}

Offset field in the rotating frame:
$$B_{off} = B_z - \omega_{TX}/\gamma$$

In the frame rotating with the transmitter frequency ω_{TX} in resonance with the Larmor frequency $\omega_0 = \gamma\,B_0$ of the homogeneous magnetic field B_0, the superimposed gradient field $B_z(r) - B_0 = B_{off}(r)$ leads to a precession of magnetization with the off-resonance frequency $\Omega(r) = \gamma\,B_{off}(r)$

Constant and Linear Magnetic Fields

- *NMR spectroscopy* requires magnetic fields sufficiently homogeneous across a typically 5-mm diameter sample tube to resolve differences in the *chemical shift* of 0.01 ppm = 10^{-8}
- *Magnetic resonance imaging* (*MRI*) does not primarily aim at resolving chemical shift differences
- The dominating signal in medical MRI results from water, which produces a single resonance line in the ^{1}H NMR spectrum
- The signals from different positions in an object are identified by means of their resonance frequencies $\omega = \gamma\,B$, provided the magnetic field $B(x,y,z)$ is different at each volume cell (*voxel*) in the object
- Along one dimension, the frequency is preferentially made to vary linearly with position r in the laboratory frame by assuring that the magnetic field scales linearly with position, for example, $B_z = B_0 + G\,r$, so that $\omega = \gamma\,(B_0 + G\,r)$
- The quantity $B_z - B_0 = B_{off} = G\,r$ is the offset-field in the coordinate frame rotating with the transmitter reference frequency $\omega_{TX} = \gamma\,B_0$
- With MRI instruments, linearly varying offset fields can be generated in all three dimensions of the laboratory frame by driving currents through *gradient coils* that produce magnetic *gradient fields* $G_x\,x$, $G_y\,y$, $G_z\,z$
- A linear variation of the magnetic field implies that the gradient is constant and not linear as often mistakenly stated in the literature

Static Magnetic Fields with Gradients

Magnetic gradient field in z direction: $\boldsymbol{B} = \begin{pmatrix} 0 \\ 0 \\ B_z \end{pmatrix}$

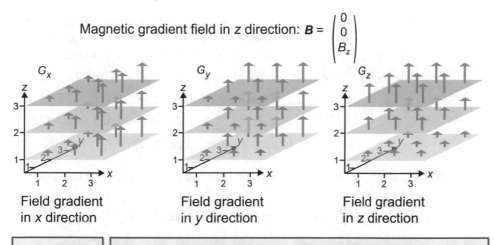

Field gradient in x direction	Field gradient in y direction	Field gradient in z direction

$\boldsymbol{G} = \begin{pmatrix} G_x \\ G_y \\ G_z \end{pmatrix}$

In high-field NMR the field gradient is approximated by a vector $\boldsymbol{G}$

In general the gradient vector $\boldsymbol{G}$ points into a direction different from the magnetic field vector $\boldsymbol{B}$

Gradient Fields and Field Gradients

- The direction of the *gradient field* generated by the *gradient coils* is aligned by default with the direction of the homogeneous magnetic field $\boldsymbol{B} = (0,0,B_0)^\dagger$
- The gradient field $B_{\text{off}} = G_x\,x + G_y\,y + G_z\,z = \boldsymbol{G}\,\boldsymbol{r}$ therefore adds to the z-component of the magnetic field producing the applied magnetic field $\boldsymbol{B} = (0,0,B_0 + \boldsymbol{G}\,\boldsymbol{r})^\dagger$
- In strict terms, the magnetic field gradient is a tensor or a 3×3 matrix in Cartesian coordinates with elements $G_{mn} = \partial B_m/\partial r_n$, and Maxwell's equations yield contributions B_x and B_y, which are due to the fact, that the magnetic field lines are curved and form closed loops
- At high magnetic field B_0, however, the B_x and B_y terms can be neglected so that in the high-field approximation the gradient is a vector $\boldsymbol{G} = (G_x, G_y, G_z)^\dagger$
- The direction of this vector is determined by the currents driving the three gradient coils that produce the gradient fields $G_x\,x$, $G_y\,y$, and $G_z\,z$, where each field is aligned with the $\boldsymbol{B}_0$ field direction
- In general, the directions of the *field-gradient vector* and the *gradient-field vector* are different and so are their units

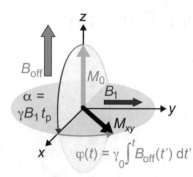

The Acquired Signal

Following a 90° pulse and neglecting chemical shift, for each volume cell (voxel) at position r the transverse magnetization is given by

$M_{xy}(r,t) = M_z(r) \exp\{-[1/T_2(r) - i \, \Omega(r)] \, t\}$, where

$\Omega = \gamma \, (B_z - B_0) = \gamma \, B_{off}$, and

$M_{xy} = M_x + i \, M_y = M_z(r) \exp\{-t/T_2(r)\} \exp\{i\varphi(r)\}$

In thermodynamic equilibrium $M_z(r) = M_0(r)$

In general, Ω depends on time. Then $\varphi = \Omega \, t$ becomes $\varphi(t) = {}_0\!\int^t \Omega(t') \, dt'$ and

$$M_{xy}(r,t) = M_z(r) \exp\{-t/T_2(r) + i \, {}_0\!\int^t \Omega(r,t') \, dt'\}$$

For non-exponential relaxation, the relaxation decay $\exp\{-t/T_2\}$ assumes the general envelope $a(t)$, and

$$M_{xy}(r,t) = M_z(r) \, a(r,t) \exp\{i \, \varphi(r,t)\} = M_z(r) \, a(r,t) \exp\{i \, {}_0\!\int^t \Omega(r,t') \, dt'\}$$

The NMR Signal in a Volume Cell

- In heterogeneous objects, the *longitudinal magnetization* M_z and the *transverse magnetization* $M_{xy} = M_x + i \, M_y \equiv M \exp\{i\varphi\}$ generated by a 90° pulse depend on the position r within the sample
- In thermodynamic equilibrium, the longitudinal magnetization $M_0(r)$ per volume element is the *spin density*
- Following an excitation pulse, the transverse magnetization $M_{xy}(r,t)$ of a voxel precesses with frequency Ω around the z axis in the *rotating coordinate frame*. Its magnitude ideally decreases exponentially with T_2
- The *precession frequency* Ω is determined by the *off-set field* in the rotating frame, which is approximated for a linear *gradient-field* by $B_{off} = B_z - B_0 = \boldsymbol{G} \, r$, where $\boldsymbol{G}$ is the *field-gradient vector*, which collects the spatial derivatives of B_z
- For spins moving from one value B_{off} of the field to another, the precession frequency changes with time. So does the *precession phase*, so that the phase is written in integral form, $\varphi(r,t) = {}_0\!\int^t \Omega(r,t') \, dt'$
- For the sake of simplicity, Ω denotes a right-handed precession
- To accommodate transverse relaxation decays other than of exponential form, a generalized signal *attenuation function* $a(r,t)$ is introduced instead of $\exp\{-t/T_2(r)\}$

The Precession Phase

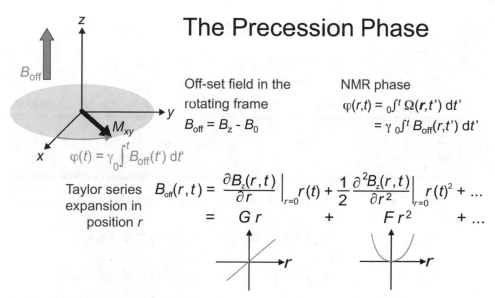

Off-set field in the rotating frame

$$B_{off} = B_z - B_0$$

NMR phase

$$\varphi(r,t) = {}_0\!\int^t \Omega(r,t')\, dt'$$
$$= \gamma\, {}_0\!\int^t B_{off}(r,t')\, dt'$$

$$\varphi(t) = \gamma\, {}_0\!\int^t B_{off}(t')\, dt'$$

Taylor series expansion in position r

$$B_{off}(r,t) = \left.\frac{\partial B_z(r,t)}{\partial r}\right|_{r=0} r(t) + \frac{1}{2}\left.\frac{\partial^2 B_z(r,t)}{\partial r^2}\right|_{r=0} r(t)^2 + \dots$$
$$= \quad G\, r \quad\quad + \quad\quad F\, r^2 \quad\quad + \dots$$

Taylor series expansion in time t

$$r(t) = r_0 + \left.\frac{\partial r}{\partial t}\right|_{t=0} t + \frac{1}{2}\left.\frac{\partial^2 r}{\partial t^2}\right|_{t=0} t^2 + \dots = r_0 + v_0 t + \frac{1}{2} a_0 t^2 + \dots$$
$$r^2(t) = r_0^2 + 2\, r_0 v_0 t + (v_0^2 + r_0 a_0)\, t^2 + \dots$$

Dependence on Time and Position

- The fundamental quantity of importance for space encoding is the *phase φ of the transverse magnetization $M_{xy}(r,t)$*
- The variation of the *magnetization phase* with position r depends on the profile of the off-resonance field across the sample. If unknown, the magnetic field is expanded into a *Taylor series*
- For simplicity of notation, only one component r of position is considered in the following. Then associated parameters including the Taylor expansion coefficients of the magnetic field profile become scalar quantities
- For conventional imaging only the linear expansion coefficient, i.e. the *gradient G* of the field profile is important. The *curvature F* is usually made as small as possible by the spectrometer hardware. But it may assume significant values in unilateral NMR and in the pores of porous media
- If the nuclear spins are moving through the sample by random *diffusion* or *coherent flow*, their positions depend on time
- For motions slow compared to the time scale of the NMR experiment the time-dependent position is expanded also into a Taylor series, which involves initial *position r_0*, initial *velocity v_0*, and initial *acceleration a_0*

Moments and Fourier Pairs

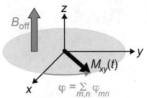

$$M_{xy}(t) = M_z(0)\,\exp\{-t/T_2 + i\,\gamma_0\int^t B_{off}(t')\,dt'\}$$
$$= M_z(0)\,\exp\{-t/T_2\}\,\exp\{i\,\Sigma\,\varphi_{mn}\}$$

$$\varphi = \sum_{m,n}\varphi_{mn}$$

Phases φ_{mn} acquired by spins moving in linear and quadratic fields

m \ n	0	1	2	
0	$\Omega_0\,t$	$\gamma_0\int^t G(t')\,t'^0\,dt'\,r_0$	$\gamma_0\int^t F(t')\,t'^0\,dt'\,r_0^2$	....
1		$\gamma_0\int^t G(t')\,t'^1\,dt'\,v_0$	$\gamma_0\int^t F(t')\,t'^1\,dt'\,2\,r_0\,v_0$	....
2		$\gamma_0\int^t G(t')\,t'^2\,dt'\,\tfrac{1}{2}\,a_0$	$\gamma_0\int^t F(t')\,t'^2\,dt'\,(v_0^2 + r_0\,a_0)$	....
⋮		⋮	⋮	....
0	$\Omega_0\,t$	$k\,r_0$	$\kappa\,r_0^2$	
1		$q_v\,v_0$	$\xi\,2\,r_0\,v_0$	
2		$\varepsilon\,a_0$	$\zeta\,(v_0^2 + r_0\,a_0)$	

Truncated Phase Evolution

- The *Taylor expansions* in position and time are inserted into the expression for the *phase of the transverse magnetization* producing a sum of phase terms
- The overall precession phase becomes a sum of terms associated with linear and curved fields as well as with different parameters of motion
- *Linear field profiles* can readily be generated with many NMR instruments
- For irregular motions, and for motions with a spectrum of correlation times, a frequency domain analysis can be developed
- The truncated expansion of the time- and position-dependent phase involves the integrals of the time-dependent gradients $G(t)$ and curvatures $F(t)$
- $G(t)$ and $F(t)$ can be manipulated during the NMR experiment. Typically $F(t) = 0$, and $G(t)$ is modulated in terms of positive and negative, ideally rectangular pulses of variable amplitude
- The gradient integrals are the different time *moments of the gradient modulation function*
- Including the gyro-magnetic ratio γ, the products of the moments and the associated kinetic variables r_0, v_0, and a_0 form individual phase contributions
- Thus, the integrals denoted by k, q_v, ε are *Fourier conjugated variables* to r_0, v_0, and a_0, i.e. together with their partner variables they form *Fourier pairs*
- From a systematic variation of k, q_v, and ε with measurements of the associated values of the phases φ, the quantities r_0, v_0, and a_0 can be determined

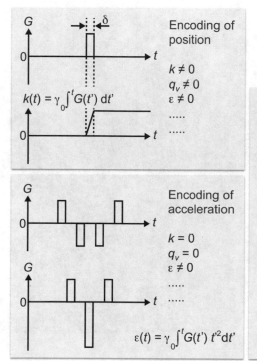

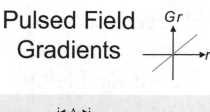

Pulsed Field Gradients

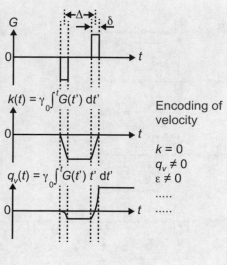

Manipulation of Gradient Moments

- *Gradient moments* are varied either by pulsing a magnetic gradient field, which is generated with current bearing coils near the sample, or by means of rf pulses
- In *pulsed field-gradient NMR (PFG NMR)* roughly rectangular gradient-field pulses are generated with durations of 10 μs to 100 ms and gradient strengths of 0.01 T/m to 10 T/m
- If the pulses are narrow, each of them generates a value for k at the particular time it is set
- On the other hand, a pair of gradient pulses with equal amplitudes but opposite phases separated by a time delay Δ generates a value for q_v while $k = 0$ at the end of the gradient-field pulse sequence
- Such a pair encodes a phase $\varphi = k\,r_2 - k\,r_1 = k\,(r_2 - r_1)$, where r_2 and r_1 are the final and initial positions of the magnetization component travelling for the time Δ in the direction of the field gradient
- Approximating velocity as $v = (r_2 - r_1)/\Delta$, the phase encoded by the anti-phase pulsed field-gradient pair becomes $\varphi = k\,\Delta\,(r_2 - r_1)/\Delta = q_v\,v$. It is identified as the phase term φ_{11} in the phase table
- Two anti-phase pairs of anti-phase gradient pulse pairs then generate a value for ε with $q_v = 0$ and $k = 0$ at the end of the sequence. Note that the two inner gradient pulses can be contracted into one to encode acceleration

Phase Encoding with Linear and Quadratic Fields

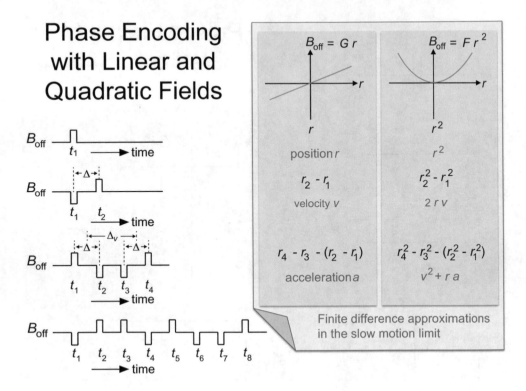

Finite difference approximations in the slow motion limit

Encoding Time Derivatives of Position

- A short pulse of a spatially *linear magnetic field* B_{off} marks *position*
- A short pulse of a *quadratic magnetic field* B_{off} marks position square
- An anti-phase pulse pair of a linear magnetic offset field B_{off} marks negative initial and positive final positions of a moving transverse magnetization component at times t_1 and t_2, respectively, separated by the pulse spacing Δ
- By dividing the marked position difference by the encoding time Δ, *velocity* is measured in a *finite difference approximation*
- Similarly, *acceleration* can be measured in a finite difference approximation
- Also other finite difference schemes known from numerical differentiation algorithms may be employed to encode these and other kinetic variables of translational motion
- This finite difference approach applies to arbitrary profiles of the offset field including the quadratic field profile

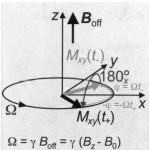

$\Omega = \gamma\, B_{off} = \gamma\, (B_z - B_0)$

A 180° pulse changes the sign of the precession phase φ. Example: $180°_x$ pulse at $t = t_E/2$:

Before:
$M_{xy}(t_-) = M_{xy}(0)\, \exp\{i\, \Omega t_-\}$
$\qquad = M_{xy}(0)\, (\cos \Omega t_-$
$\qquad\qquad + i \sin \Omega t_-)$

After:
$M_{xy}(t_+) = M_{xy}(0)\, (\cos \Omega t_+$
$\qquad\qquad - i \sin \Omega t_+)$
$\qquad = M_{xy}(0)\, \exp\{-i\, \Omega t_+\}$

The Effective Gradient

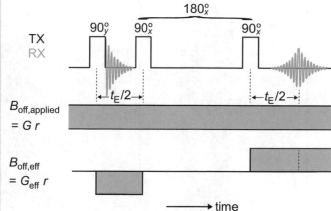

In the high-field approximation the spatial dependence of the off-resonance field B_{off} affects only transverse magnetization. This is used to generate pulses of effective gradients G_{eff} via rf pulses in time-invariant gradient fields $B_{off} = G\, r$

Time-Invariant Offset Fields

- In the magnetic resonance community, *pulsed field-gradient NMR (PFG NMR)* refers to NMR with *pulsed gradient fields* $B_{off} = G\, r$
- In pulse-sequence diagrams it is common practice to plot the time-dependence of the *field gradient G* and not of the applied *gradient field* $G\, r$
- The time-dependence of any off-set field affects only coherences but not the longitudinal magnetization as long as the offset field $B_{off} = G\, r$ is weak compared to the polarization field B_0
- Then the offset field affects the measured signal only during the evolution and detection periods, for example, when transverse magnetization evolves
- In consequence, time-invariant offset fields appear as pulses of an effective offset field B_{off} affecting transverse magnetization, for example, in the *stimulated echo* sequence
- For time-invariant linear offset fields $B_{off} = G\, r$ the time-dependent *effective gradient* G_{eff} is introduced
- Note, that the sign of any effective gradient in the past changes with each 180° rf pulse, because with each pulse the phase of the transverse magnetization changes sign
- In the maximum of the stimulated echo the first-order moment of the effective gradient modulation function $_0\!\int^{tE} G_{eff}(t)\, dt = k(t_E)/\gamma = 0$, which is the condition for a *gradient echo*

Transverse Magnetization in a Gradient Field

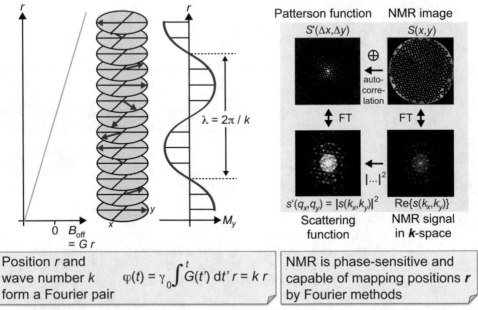

From B. Blümich, *NMR Imaging of Materials*, **Clarendon Press**, Oxford, 2000, Figs. 2.2.4, 5.4.3 by permission of Oxford University Press

Interpretation of the Gradient Integral

- In a magnetic field, which varies linearly with position r, the spins precess with linearly varying *Larmor frequency*
- Following a gradient-field pulse the magnetization phase in adjacent voxels along the gradient direction then varies linearly with position
- The tips of the magnetization vectors lie on a helix, which winds around the direction r of the magnetic field gradient
- The projections of the magnetization vectors onto a transverse axis in the RCF produces a sinusoidal function with period $\lambda = 2\pi/k$ and phase $\varphi = k\,r$
- Therefore, k is the *wave number*, which specifies the spatial oscillation frequency of the transverse magnetization in analogy to ω, which specifies the temporal oscillation frequency
- An NMR image derives by Fourier transformation (FT) from transverse magnetization $M_{xy}(k_x, k_y) = s(k_x, k_y)$ acquired for a range of wave numbers along both image dimensions
- If the object has periodic structure, $s(k_x, k_y)$ shows *interference fringes*
- By forming the magnitude square of $s(k_x, k_y)$, the *scattering function* is obtained. The NMR phase is discarded but the interference fringes remain
- The Fourier transform of the scattering function is known as the *Patterson function*. It is the *auto-correlation function* of the image
- The Patterson function and $|s(k_x, k_y)|^2$ no longer give access to positions x, y but only to position differences Δx, Δy in analogy to X-ray *diffraction* data
- $\Delta x, \Delta y = \Delta r$ derive from the wave numbers $q_x, q_y = q$ of the peaks in the scattering function, because Δr and q form a Fourier pair corresponding to $\varphi = q\,\Delta r$

Transverse Magnetization in Linear Fields

For simplicity of notation the indices '0' with r_0, v_0, a_0 are dropped form now on. Consider the signal from a single voxel at position r:

$$M_{xy}(t, \varphi) = M_{xy0}(r,v)\, a(t,r)\, \exp\{i\; \varphi(t)\}$$
$$= M_{xy0}(r,v)\, a(t,r)\, \exp\{i\; [\Omega(r)t + k(t)r + q_v(t)v(r) + ...]\}$$

The signal from the whole sample is the integral over all voxels:

$$M_{xy}(t,k,q_v) = \iiint M_{xy0}(\Omega,r,v)\, a(t,r)\, \exp\{i\; [\Omega(r)t + k(t)r + q_v(t)v(r) + ...]\}\; d\Omega\, dr\, dv$$

Spectroscopy: $k = 0 = q_v$

$$M_{xy}(t) = \iiint M_{xy0}(\Omega,r,v)\, a(t,r)\, \exp\{i\; \Omega(r)\, t\}\; d\Omega\, dr\, dv$$

Imaging and flow: $t_G \ll$ correlation time of motion; no dispersion in frequency Ω

Imaging with *phase encoding*: $t = t_E$, $k = \gamma\, G\, t_E$, $q_v = 0$. Fix t_E, vary k,
$$M_{xy}(k) = \iiint M_{xy0}(r,v)\, a(t_E,r)\, \exp\{i\; k(t_E)\, r\}\; d\Omega\, dr\, dv$$
Imaging with *frequency encoding*: $k = \gamma\, G\, t$, $q_v = 0$. Vary t and k,
$$M_{xy}(t) = \iiint M_{xy0}(r,v)\, a(t,r)\, \exp\{i\; \gamma\, G\, r\, t\}\; d\Omega\, dr\, dv$$
Velocity distributions by phase encoding: $t = t_E$, $k = 0$. Fix t_E, vary q_v,
$$M_{xy}(q_v) = \iiint M_{xy0}(r,v)\, a(t_E,r)\, \exp\{i\; [q_v(t_E)\, v(r)]\}\; d\Omega\, dr\, dv$$
Flow imaging by phase encoding: $t = t_E$. Fix t_E, vary k and q_v,
$$M_{xy}(k,q_v) = \iiint M_{xy0}(r,v)\, a(t_E,r)\, \exp\{i\; [k(t_E)\, r + q_v(t_E)\, v(r)]\}\; d\Omega\, dr\, dv$$

Information Accessible by PFG NMR

- Gradient fields are applied for a time t_G to identify the NMR signal from different voxels along the gradient direction
- The signal acquired from a heterogeneous sample is the integral of the signal from all voxels at positions r
- For spins in motion, one also needs to integrate over all velocities v
- In general the signal is acquired as a function of k and q_v, where both variables depend on time t via the *moments of the gradient modulation function*
- The measured signal $M_{xy}(t,k,q_v)$ derives from the longitudinal magnetization $M_z(r_0,v_0)$ following a 90° pulse
- Apart from the attenuation function $a(t,r_0)$, $M_{xy}(r_0,v_0)$ is the *Fourier transform of the measured signal*
- Therefore, the signal is acquired for a sufficiently large number of values k and q_v and is subsequently Fourier transformed
- The *attenuation function* $a(t,r_0)$ introduces a loss of image resolution, because the NMR image is the convolution of $M_z(r_0,v_0)$ with the Fourier transform of $a(t,r_0)$
- The Fourier transform of $a(t,r_0)$ is referred to as the *point-spread function*. It specifies the spatial resolution of the image
- When starting the imaging experiment from thermal equilibrium, $M_z(r_0,v_0) = M_0(r_0,v_0)$ is proportional to the *spin density* of the object

Sampling of k Space

Cylindrical coordinates: Back-projection imaging: $\tan \phi = G_y/G_x$

Data sampled at discrete points in time lead to discrete points in k-space

Cartesian coordinates: Conventional Fourier imaging

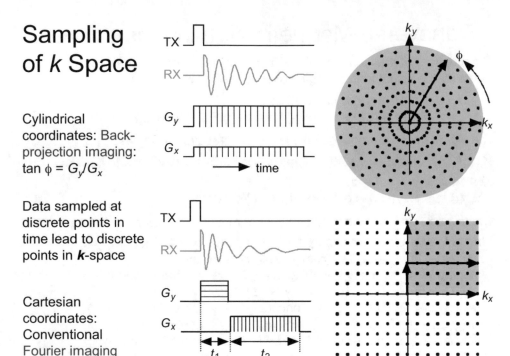

Principles of 2D Imaging

- With *Fourier NMR*, images are measured by acquiring the complex transverse magnetization $M_{xy} = M_x + iM_y$ as a function of $k = (k_x, k_y, k_z)^\dagger$ and Fourier transforming the measured data over k
- The values of k are scanned on a discrete grid in one, two, or three dimensions
- Historically, two principle approaches are discriminated: The k-space grid can be defined on spherical/*cylindrical coordinates* and on *Cartesian coordinates*
- These schemes are commonly known as *back-projection* (BP) *imaging* and as *Fourier imaging*, although Fourier transformation is required in both cases
- The space-encoded data in one dimension are usually acquired directly in the presence of a time-invariant field gradient
- In BP imaging this scheme is used with repeated acquisitions under different gradient directions
- In Fourier imaging, the data for further space dimensions are acquired indirectly by pulsing gradient fields in orthogonal directions in *preparation periods* prior to data acquisition
- The Fourier transformation of BP imaging data involves the transformation from cylindrical to Cartesian coordinates because a fast *multi-dimensional Fourier transformation* is executed in Cartesian coordinates

Frequency and Phase Encoding

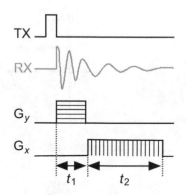
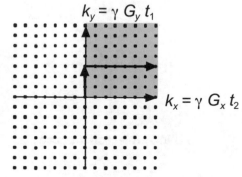

Frequency encoding: vary t_2 in n_2 steps

$M_{xy}(n_2) = \exp\{-[1/T_2^* - i \gamma G_x x] n_2 \Delta t_2\}$

Spatial resolution limited by $\Delta\omega_{1/2}$

$\gamma G_x \Delta x \geq 2/T_2^* = \Delta\omega_{1/2}$

$1/\Delta x \leq \gamma G_x T_2^* / 2 = \gamma G_x / \Delta\omega_{1/2}$

Phase encoding: vary G_y in n_1 steps

$M_{xy}(n_1) = \exp\{-[1/T_2^* - i \gamma n_1 \Delta G_y y] t_1\}$

Spatial resolution increases with $n_{1,max}$

$\gamma n_{1,max} \Delta G_y \Delta y t_1 \leq 2\pi$

$1/\Delta y \geq \gamma n_{1,max} \Delta G_y t_1 / 2\pi = k_{y,max}/(2\pi)$

Space Encoding and Resolution

- The terminology of *phase encoding* and *frequency encoding* of the space information is historic
- The acquisition of the NMR signal in the presence of a time invariant field gradient is referred to as *frequency encoding* of the space information
- In frequency encoding k_x increases with the acquisition time t_2
- With increasing t_2 not only are higher values of k being encoded but also the signal attenuation by T_2^* increases, so that the *spatial resolution* $1/\Delta x$ is limited by the line width $\Delta\omega_{1/2} = 2/T_2^*$
- Modulation of the initial phase of the acquired magnetization in an *evolution time* t_1 prior to the *acquisition time* t_2 is referred to as *phase encoding*
- In phase encoding k_y is varied preferably by changing the gradient amplitude in steps ΔG instead of the gradient duration in steps Δt. This method is known as *spin-warp imaging*. It avoids variable signal attenuation from T_2 relaxation and signal modulations by the chemical shift and other spin interactions
- In spin-warp phase encoding the spatial resolution $1/\Delta y$ is limited by the maximum gradient strength $n_{1max} \Delta G_y$
- Conventional 2D and 3D echo imaging methods combine spin-warp phase encoding and frequency encoding
- Pure phase encoding is used for *spectroscopic imaging*, *velocity encoding*, and for *imaging of solids*. It is referred to also as *single-point imaging* (*SPI*)

Scanning 2D **k** Space

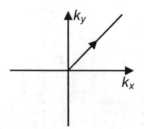

Back-projection imaging

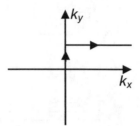

Spin-echo imaging

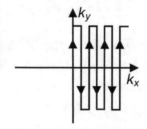

Echo planar imaging (EPI)

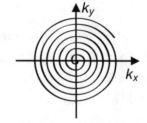

Spiral imaging

Walking Through **k** Space

- The information depicted in an NMR image centers at the **k**-space origin
- To acquire the NMR signal with spatial resolution, it needs to be acquired for a region of **k** space centered at **k** = 0
- The signal in this region is defined on a discrete grid of points
- The sequence in which these points are addressed is determined in the imaging experiment
- The signal of a group of points is usually measured in one scan, which often includes the origin of one of the components of the **k** vector
- Typically many scans are needed to cover a complete region of **k** space
- Some *fast imaging* methods cover all relevant points of **k** space in one scan
- By *line-scan imaging* methods, the data from one line usually passing through the origin of **k** space are measured in one scan
- *Back-projection imaging* relies on several line-scans from different gradient directions
- *Spin-echo imaging* acquires data from parallel lines in subsequent scans
- *Echo planar imaging* (*EPI*) acquires the data of an entire image in a single shot following different traces through **k** space, e.g. meander and spiral traces
- *Sparse sampling* skips certain regions in **k** space to accelerate the data acquisition on the expense of image quality

Slice Selection

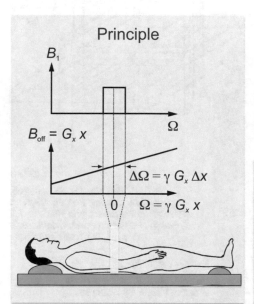

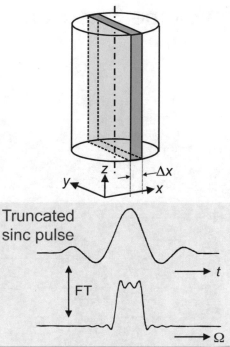

From P. G. Morris, *Nuclear Magnetic Resonance Imaging in Medicine and Biology*, **Clarendon Press**, Oxford, 1986, Fig. 3.8 by permission of Oxford University Press

Making 3D Objects Appear Like 2D Objects

- To make 3D objects appear two dimensional in NMR, a *projection* must be measured or the magnetization of a 2D *slice* through the object
- Frequently images of 2D slices through 3D objects are measured
- To select a slice, a *frequency-selective pulse* is applied with the object exposed to an inhomogeneous magnetic field, which typically is a linear field with a space invariant gradient
- The linear field identifies different positions along the gradient direction by their NMR frequencies
- Constant frequencies are found in planes orthogonal to the gradient direction
- The frequency-selective pulse acts on the magnetization components within a limited frequency region only
- The width of the frequency region defines the thickness of the selected slice
- To a first approximation, the *Fourier transform* of the time-domain pulse shape defines the frequency-selection properties of the pulse
- A pulse with a *sinc* shape in the time domain exhibits a rectangular profile in the frequency domain
- To obtain pulses with finite durations, the lobes in the time domain are truncated on the expense of perfect *slice definition* in the frequency domain

Volume Selection

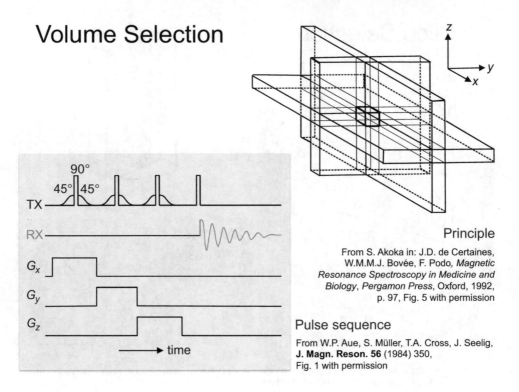

Principle

From S. Akoka in: J.D. de Certaines,
W.M.M.J. Bovée, F. Podo, *Magnetic
Resonance Spectroscopy in Medicine and
Biology, Pergamon Press*, Oxford, 1992,
p. 97, Fig. 5 with permission

Pulse sequence

From W.P. Aue, S. Müller, T.A. Cross, J. Seelig,
J. Magn. Reson. 56 (1984) 350,
Fig. 1 with permission

Restricting the Signal-Bearing Volume

- To investigate small regions within large objects by NMR imaging or to measure NMR spectra from well defined regions within the object (*volume-selective spectroscopy*), the magnetization within the selected volume must be accessed without crosstalk from magnetization in the surrounding volume
- Preferentially *longitudinal magnetization* is prepared because it relaxes slower than *transverse magnetization* in inhomogeneous fields
- Positions within the sample are labeled using magnetic field gradients in the same way as for slice selection of transverse magnetization
- The longitudinal magnetization outside the selected volume is eliminated
- The sensitive volume is defined at the intersection of three orthogonal *slices*
- Each pulse for selection of longitudinal magnetization consists of a package of three pulses, a selective 45° pulse, a nonselective 90° pulse, and another selective 45° pulse
- The first 45° pulse tips the magnetization within the selected plane by 45°
- The nonselective 90° pulse rotates the complete magnetization of the sample by 90°
- The last selective 45° pulse continues to rotate the magnetization of the slice through another 45°. In the end the magnetization of the slice has been rotated through a total angle of 180° and ends up as longitudinal magnetization
- The unwanted magnetization has been rotated by 90° only and dephases as transverse magnetization

Spin-Echo Imaging

(Spin-warp imaging)

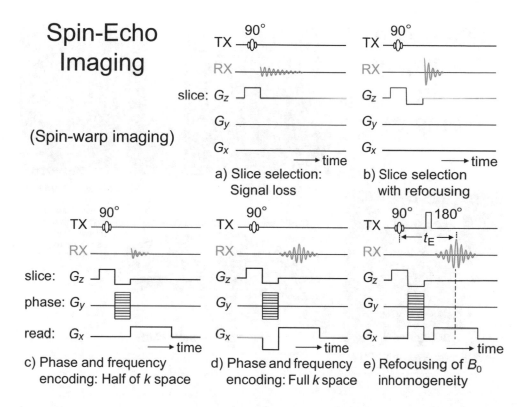

a) Slice selection: Signal loss

b) Slice selection with refocusing

c) Phase and frequency encoding: Half of k space

d) Phase and frequency encoding: Full k space

e) Refocusing of B_0 inhomogeneity

Slice-Selective 2D Spin-Echo Imaging

- A 2D imaging scheme starts by preparing the magnetization in a 2D *slice* selected with a suitably shaped pulse applied in the presence of a field gradient (a)
- The *slice-selective pulse* is long, and the magnetization dephases during the pulse in the inhomogeneous field
- This dephasing is refocused in a *gradient echo* generated by a second, negative field gradient pulse with an area half of that of the first gradient pulse (b)
- The 2D space information is inscribed into the *transverse magnetization* of the selected slice by *phase encoding* in an *evolution period* and by *frequency encoding* in the *detection period* (c)
- The *gradient switching times* are finite, and some signal is lost during these times. This signal needs to be recovered
- Also, components at negative k need to be encoded during the detection time
- Both demands are met by forming a *gradient echo* during the detection time (d)
- This echo is generated by starting the frequency-encoding gradient with an initial negative lobe having an area half of that of the succeeding positive lobe
- Signal dephasing from chemical shift distributions and inhomogeneity of the detection field can in addition be refocused in a *Hahn echo* by applying a 180° pulse separating evolution and detection periods (e)
- Then, the signs of the field-gradient pulses applied during the evolution time need to be inverted
- This method of 2D imaging is known as *spin-echo imaging*

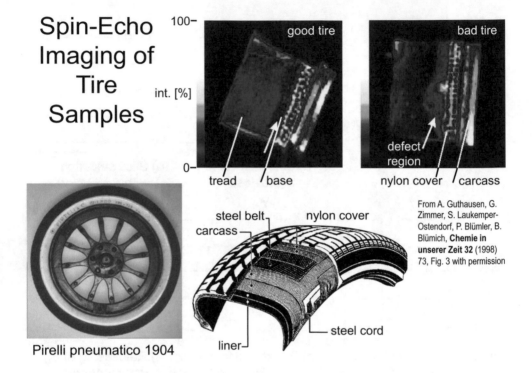

From A. Guthausen, G. Zimmer, S. Laukemper-Ostendorf, P. Blümler, B. Blümich, **Chemie in unserer Zeit 32** (1998) 73, Fig. 3 with permission

An Example of 2D Spin-Echo Imaging

- 2D *spin-echo imaging* used to be applied in clinical MRI. But imaging times are long, because the longitudinal magnetization needs to recover after each scan in preparation times approximately 5 T_1 long
- *Rubber* is an inhomogeneous product especially when filled with carbon black. Consequently signal lost from magnetization dephasing in local field distortions originating from changes of the magnetic susceptibility within the sample needs to be recovered in a *spin echo*
- Spin-echo imaging used to be employed in many tire development centers to probe the *vulcanization* state of different rubber layers in sample sections cut from test *tires*
- The *image contrast* is largely defined by differences in *transverse relaxation* during the *echo time* and by *longitudinal relaxation* during the *recovery period*
- Sample regions with soft rubber and mobile additives have long T_2, hard rubber has shorter T_2, and solid polymers like textile fibers have even shorter T_2
- The typical *spatial resolution* in such images is 1/(0.1 mm) in both dimensions
- The imaged slices are usually much thicker than 0.1 mm, of the order of 5 mm, in order to gain enough signal

3D Spin-Echo Imaging

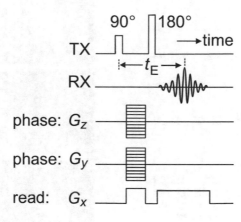

3D Imaging

- *3D images* can be obtained in different ways
- Successive 2D slices measured by *slice-selective 2D imaging* can be combined into a 3D image
- Here, the spatial resolution perpendicular to the slice plane is low. During acquisition, the signal comes from one selected slice only, while the noise comes from the entire sample
- In *3D Fourier imaging*, the signal is acquired at all times from the same volume, which also produces the noise, and the resolution can be high in all three dimensions
- For 3D Fourier imaging the slice-selective pulse from 2D MRI is replaced by a non-selective pulse
- A further *phase-encoding gradient* pulse is introduced and stepped through positive and negative values independent of the other gradient pulses
- The image is obtained by *3D Fourier transformation* of the acquired data set

Gradient-Echo Imaging

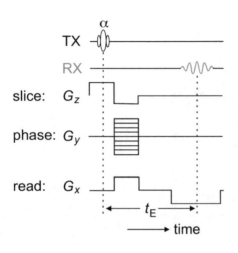

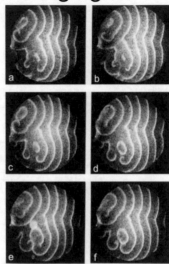

FLASH: Fast low-angle shot

From A. Haase, J. Frahm, D. Matthei, W. Hänicke, K. D. Merboldt, **J. Magn. Reson. 67** (1986) 258, Fig. 1 with permission

Mn catalyzed Belousov-Zhabotinsky reaction: Images taken at 40 s intervals

From A. Tzalmona, R. L. Armstrong, M. Menzinger, A. Cross, C. Lemaire, **Chem. Phys. Lett. 174** (1990) 199, Fig. 1 with permission

Reducing the Measurement Time

- The measurement time for a *spin-echo image* is determined by the number of lines in the image data matrix and the duration of the *recycle delay* to recover the longitudinal magnetization
- Due to the fact that in a spin echo the entire longitudinal magnetization is perturbed, the recycle delay is of the order of 5 T_1
- Shorter recycle delays can be employed if the longitudinal magnetization is only slightly perturbed from its thermodynamic equilibrium value
- This is achieved by discarding the 180° refocusing pulse in the imaging sequence. In particular, if a *small flip-angle pulse* is used instead of the initial 90° pulse to rotate the longitudinal magnetization only partially into the transverse plane, the recycle delay can be shortened considerably
- Following Richard Ernst, the optimum flip angle α_E and the optimum recycle delay t_0 are determined by the longitudinal relaxation time T_1 according to $\cos\alpha_E = \exp\{-t_0/T_1\}$, where α_E is the *Ernst angle*
- The resultant imaging method is called *gradient echo imaging* or fast low-angle shot (*FLASH*) *imaging*
- It is a standard method in *clinical imaging*
- It has also been used for imaging slow dynamic processes in soft matter, such as chemical waves in oscillating reactions

Susceptibility Contrast

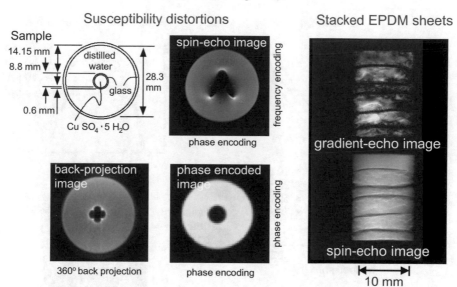

Susceptibility distortions

Sample
14.15 mm
8.8 mm
distilled water
28.3 mm
glass
0.6 mm
Cu SO$_4$ · 5 H$_2$O

spin-echo image

frequency encoding

phase encoding

back-projection image

phase encoded image

360° back projection

phase encoding

phase encoding

From O. Beuf, A. Briguet, M. Lissac, R. Davis, **J. Magn. Reson. B 112** (1996) 111, Figs. 2, 5 with permission

Stacked EPDM sheets

gradient-echo image

spin-echo image

10 mm

From P. Blümler, V. Litvinov, H.G. Dikland, M. van Duin, **Kautschuk Gummi Kunststoffe 51** (1998) 865, Fig. 1 with permission

Contrast and Artifacts

- In spin-echo imaging, the image *contrast* is determined by three factors
- 1) The *spin density*, i.e. the number of nuclei at position r contributes $M_{z0}(r)$
- 2) *Transverse relaxation* during the echo time t_E contributes $\exp\{-t_E/T_2(r)\}$
- 3) *Partial saturation* due to short recycle delays t_0 contributes $1-\exp\{-t_0/T_1(r)\}$
- In total, the image amplitude is given by $M_{z0}(r) \exp\{-t_E/T_2(r)\} [1-\exp\{-t_0/T_1(r)\}]$
- In addition, the contrast is enhanced by *magnetic field distortions* in sample regions, where the *magnetic susceptibility* changes
- Examples are carbon-black filler clusters embedded in a rubber matrix and the glass interface between distilled water and water doped with copper sulfate
- In *gradient-echo imaging*, the magnetization dephasing due to *magnetic field inhomogeneity is* not refocused, so that susceptibility distortions are enhanced in the frequency-encoding direction
- These distortions do not appear in the *phase-encoding dimension* because the encoding time t_1 is kept constant
- In *back-projection imaging* with frequency encoding in both dimensions, the artifacts appear symmetrically in both dimensions
- In pure *phase-encoding imaging*, they are not observed at all
- In standard spin-echo imaging, the dephasing from magnetic field inhomogeneity is refocused for the phase-encoding dimension in the echo maximum
- *Susceptibility effects* can be considered artifacts in images, but can also be used to generate image contrast, e.g. in carbon-black filled *elastomers*

Temperature Imaging from Relaxation Maps

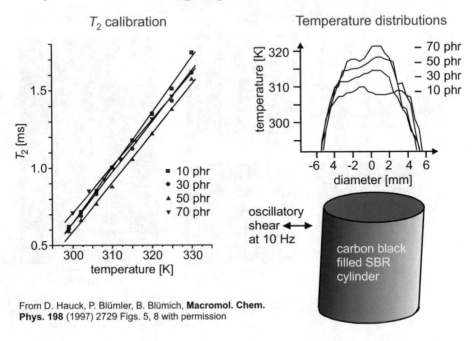

T_2 calibration

Temperature distributions

- 10 phr
- 30 phr
- 50 phr
- 70 phr

— 70 phr
— 50 phr
— 30 phr
— 10 phr

oscillatory shear ⟷ at 10 Hz

carbon black filled SBR cylinder

From D. Hauck, P. Blümler, B. Blümich, **Macromol. Chem. Phys. 198** (1997) 2729 Figs. 5, 8 with permission

Parameter Images

- Images of the spin density $M_{z0}(r)$, which are weighted by a function of other NMR parameters, for example by $\exp\{-t_E/T_2(r)\}$, are called *parameter-weighted spin-density images*
- By acquisition of several images with different echo times, the parameter $T_2(r)$ can be extracted for every voxel at position r from the set of images
- The resultant map of $T_2(r)$ is called a T_2 *image* or in general a *parameter image*
- In elastomers, T_2 is often found to be proportional to temperature within small temperature ranges
- Then, a T_2 parameter image $T_2(r)$ can be calibrated into a *temperature map*
- Such a temperature map has been determined by NMR for a carbon-black filled rubber cylinder undergoing small oscillatory shear deformation at a frequency of 10 Hz
- Due to the dynamic loss modulus, some deformation energy is dissipated as thermal energy
- The heat gain inside the sample competes with the heat loss through the sample surfaces
- The resultant *temperature distribution* exhibits a peak in the center of the sample
- Dynamic mechanical load thus leads to dynamic sample heterogeneity

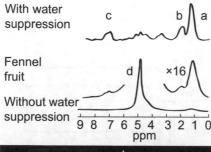

With water
suppression

Fennel
fruit

Without water
suppression

Spectroscopic
Spin-Echo Imaging

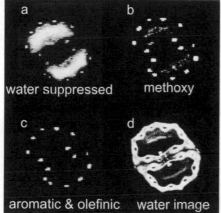

water suppressed   methoxy

aromatic & olefinic   water image

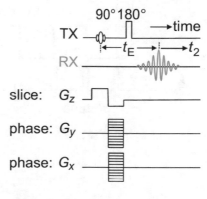

From H. Rumpel, J. M. Pope, **Magn. Reson.**
Imag. 10 (1992) 187, Fig. 8 with permission

Incorporating a Spectroscopic Dimension

- An *NMR spectrum* can be acquired in each voxel either indirectly or directly
- Indirect acquisition of the spectrum demands stepping point by point through the FID in an evolution time with different scans while the field gradient is off
- Direct acquisition corresponds to sampling the FID in a homogeneous field during the detection period
- The typical *digital resolution* of an NMR image is 256 points in each dimension
- A 1D NMR spectrum consists of roughly 1024 = 1k to 64 k data points
- On account of short measuring times, the spatial information is encoded for all dimensions indirectly in the signal phase during an evolution period t_1, and the spectroscopic information is acquired directly during the detection period t_2
- From such *spectroscopic images*, other images can be derived, where the *contrast* is defined by the integral of a given line in the NMR spectrum
- In this way, the concentration distributions of different chemical compounds can be imaged, for example, in plants, muscles, and the human brain

Echo-Planar Imaging (EPI)

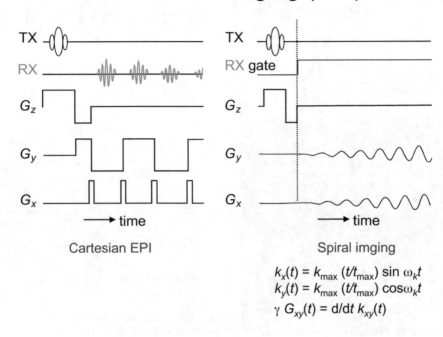

Cartesian EPI   Spiral imging

$$k_x(t) = k_{max} \, (t/t_{max}) \sin \omega_k t$$
$$k_y(t) = k_{max} \, (t/t_{max}) \cos \omega_k t$$
$$\gamma \, G_{xy}(t) = d/dt \, k_{xy}(t)$$

Acquiring Images in a Single Scan

- Fast imaging methods can be designed in different ways
- In the *steady-state free precession* method the transverse magnetization stays in dynamic equilibrium with the excitation and with relaxation
- Other methods use multiple scans in rapid succession with short or without recovery time by avoiding Hahn echoes in favor of gradient echoes. One such method is the *FLASH imaging* method
- But an entire 2D image can also be acquired in one shot by generating multiple echoes, where each echo encodes a different trace through *k*-space
- Methods of this type are referred to as *echo planar imaging* (*EPI*)
- In clinical imaging, they are usually pure gradient-echo methods to avoid excessive rf exposure
- In the original EPI method, *k*-space is scanned in either a zig-zag trace or in parallel traces on a Cartesian grid by rapidly pulsing field gradients
- When the gradient echoes are replaced by Hahn echoes, the *RARE* (Rapid Acquisition with Relaxation Enhancement) method is obtained
- A method less demanding on hardware and with less acoustic noise is *spiral imaging*, where the trace through *k*-space forms a spiral

Single-Point Imaging (SPI)

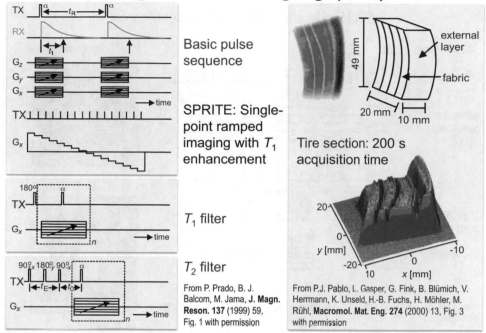

Basic pulse sequence

SPRITE: Single-point ramped imaging with T_1 enhancement

T_1 filter

T_2 filter

From P. Prado, B. J. Balcom, M. Jama, **J. Magn. Reson. 137** (1999) 59, Fig. 1 with permission

Tire section: 200 s acquisition time

From P.J. Pablo, L. Gasper, G. Fink, B. Blümich, V. Herrmann, K. Unseld, H.-B. Fuchs, H. Möhler, M. Rühl, **Macromol. Mat. Eng. 274** (2000) 13, Fig. 3 with permission

Imaging with Pure Phase Encoding

- The most successful technique for imaging objects with short transverse relaxation times such as solids or fluid-filled porous media is *single-point imaging* (SPI)
- *k*-space is sampled by pure phase encoding
- A short rf pulse with a small flip angle is applied in the presence of a field gradient
- A single point of the FID is sampled after a short dead time t_1
- The gradient is ramped to a different value and the experiment is repeated
- Depending on the flip angle and the repetition time, T_1 contrast is generated
- Different *magnetization filters* can be introduced to generate contrast by preparing the initial longitudinal magnetization in a particular way. Examples for such filters are the inversion-recovery sequence for T_1 contrast and the spin-echo sequence for T_2 contrast
- Despite the pure phase-encoding procedure, images can be acquired rather fast, such as 50 s for a single scan of an image with 128 × 64 data points

Imaging of Double-Quantum Coherences

Application to strained rubber bands with a cut

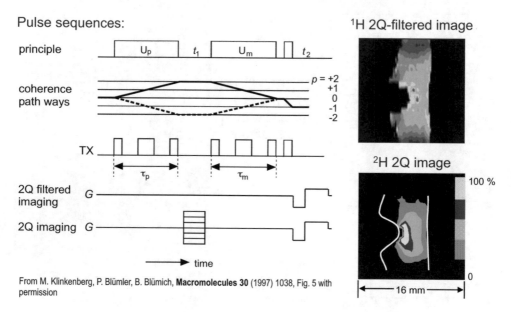

Pulse sequences:

^{1}H 2Q-filtered image

^{2}H 2Q image

From M. Klinkenberg, P. Blümler, B. Blümich, **Macromolecules 30** (1997) 1038, Fig. 5 with permission

Tricks with Coupled Spins

- Systems of coupled spins ½ and *quadrupolar spins* give rise to *multi-quantum coherences*
- The dephasing angle of coherences by precession in magnetic fields is proportional to the *coherence order p*
- *Double-quantum coherence* ($p = 2$) dephases twice as fast as *single-quantum coherence* ($p = 1$)
- Phase encoding of multi-quantum coherences effectively multiplies the strength of the field gradient with the coherence order p
- This can be made use of for imaging if the multi-quantum relaxation time is long enough, as in the case of the double-quantum coherence of deuterated oligomers incorporated into elastomers
- In other cases, the multi-quantum coherence can still be exploited in a magnetization filter preceding the image acquisition to separate signals from uncoupled and coupled spins or from isotropic and anisotropic material regions
- *Double-quantum imaging* and *double-quantum filtered imaging* have been applied to image local stress and strain in strained rubber bands by contrasting the signals from differently deformed regions of the rubber network

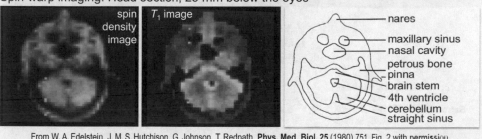

FONAR and Spin Warp Imaging

R. Damadian, L. Minkoff, M. Goldsmith, **Physiol. Chem. & Physics 10** (1978) 561, Fig. 1, private communication, 15 August 1979

From R. Damadian, L. Minkoff, M. Goldsmith, in: R. Damadian, ed., **NMR 19** (1981) 1, Fig. 10 with permission

B_z field profile

FONAR: Field Focused Nuclear Magnetic Resonance — sensitive region

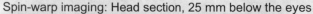

Spin-warp imaging: Head section, 25 mm below the eyes

spin density image

T_1 image

nares
maxillary sinus
nasal cavity
petrous bone
pinna
brain stem
4th ventricle
cerebellum
straight sinus

From W. A. Edelstein, J. M. S. Hutchison, G. Johnson, T. Redpath, **Phys. Med. Biol. 25** (1980) 751, Fig. 2 with permission

Early Medical Images

- The first medical images were measured in 1978 by Raymond Damadian and co-workers with the *FONAR* method
- They acquired images point-by-point in real space by shifting the patient through the sweet spot of an inhomogeneous field
- The sweet spot was generated by a saddle point in the B_0 field profile
- Within the sweet spot the field is less inhomogeneous than outside, so that the NMR signal from this region in space decays not as fast as outside
- The acquisition time was several hours, and the image quality was low
- Yet, 10 years later NMR imaging was already an indispensable diagnostic tool in hospitals all around the world
- However, it was not the FONAR method reaching the finishing line, but the *spin-warp imaging* technology developed at General Electric by W. Edelstein and collaborators
- Spin-warp imaging is spin-echo imaging with a fixed evolution time in which the gradients and not the evolution time t_1 are incremented to scan k-space
- Already the original spin-warp publication reported promising *parameter images* of the spin density and of T_1 from slices through different parts of the human body

Flow Imaging
of Blood

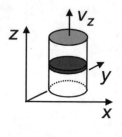

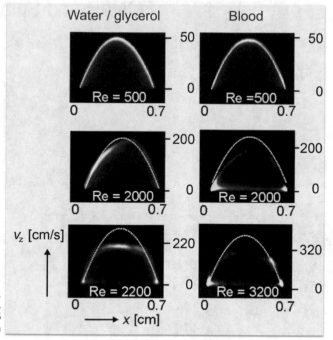

From S. Han, O. Marseille, C.
Gehlen, B. Blümich, **J. Magn.**
Reson. 152 (2001) 87, Figs. 3, 5
with permission

Imaging Flowing Liquids

- The most important application areas of non-medical imaging are in *chemical engineering* and *materials science*
- An important application field is the characterization of *transport phenomena* like fluid and granular flow or molecular self- and inter-diffusion in technical devices and porous media, which are optically non-transparent
- As long as the set-up is non-magnetic and radio-frequency transparent, the transport properties can be measured by NMR
- Usually, the phenomenon in question has to be reproduced inside the magnet unless stray-field NMR techniques are employed and the object is investigated from one side
- An important, optically opaque fluid is blood. Its rheological properties are of interest in medical engineering for building devices like artificial arteries and veins, heart valves, hemodialyzers, and blood pumps. Only the blood substitute water/glycerol is sufficiently transparent for optical velocity analysis
- With pulsed field-gradient NMR, *velocity-vector fields* can be imaged
- For quantitative MRI a slice of the moving fluid is selected, which stays inside the resonator for the durations of the space and velocity encoding periods as well as the detection period
- A complete velocity image has six dimensions: 3 for space and 3 for velocity
- Also, the velocity distribution can be determined in each voxel

Spin-Echo Pulse Sequence for Flow Imaging

Slice selection and phase encoding of v_z

Phase encoding of y

Frequency encoding of x

Evolution of gradient moments of orders 0 and 1 for position and velocity encoding in x, y and z directions

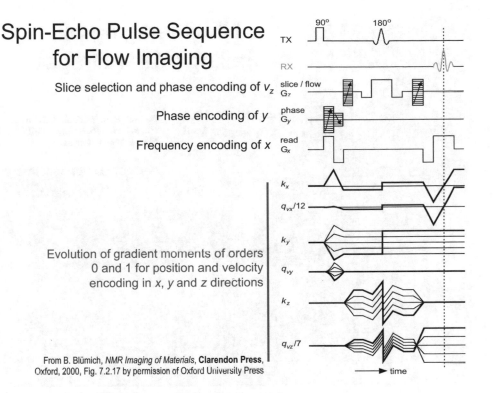

From B. Blümich, *NMR Imaging of Materials*, **Clarendon Press**, Oxford, 2000, Fig. 7.2.17 by permission of Oxford University Press

Pulse-Sequence Design

- Pulse sequences for *velocity imaging* have to incorporate *space encoding* by scanning **k** space and *flow encoding* by scanning $\boldsymbol{q}_v$ space
- The **k** and $\boldsymbol{q}_v$ spaces must be encoded independently
- For example, one **k** component is frequency encoded for direct acquisition and the other components of **k** and $\boldsymbol{q}_v$ are phase encoded in the acquired signal
- A compromise has to be made for the velocity component in direction of the frequency encoded space component. Here, complete decoupling of phase evolutions is not possible
- For example, to image $v_z(x,y)$, an xy slice is selected, and the transverse magnetization is acquired as a function of k_x, k_y, and q_{vz}
- The components q_{vx}, q_{vy}, and k_z should be zero during data acquisition
- If frequency encoding is used for k_x, then q_{vx} also varies during the detection time
- The gradient-modulation sequences for such imaging modalities are numerically optimized
- For accelerated flow the acceleration phase needs to be considered as well
- After Fourier transformation over k_x, k_y, and q_{vz}, the relaxation-weighted spin density $M_z(x,y,v_z)$ is obtained from which $v_z(x)$ and $v_z(y)$ can be extracted

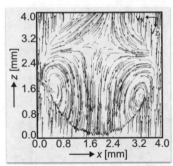

NMR Imaging of
Falling Water Drops

zx projection of the
velocity vector field

From S. Han, S. Stapf, B. Blümich,
Phys. Rev. Lett. 87 (2001) 145501,
Figs. 2, 3 with permission

Velocity-component images

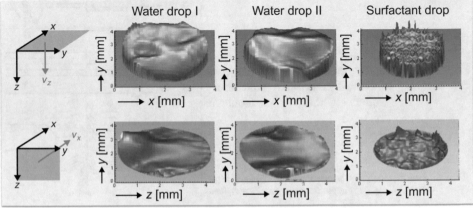

Imaging Velocity Fields

- Straight forward *flow encoding* in MRI requires long acquisition times ranging from several minutes to several hours
- Fast flow imaging schemes use echo trains to encode different components of k and q_v in each echo time and employ sparse sampling of the k and q_v spaces
- Due to long acquisition times, most flow processes suitable to conventional NMR imaging are either stationary or repetitive
- An example of a repetitive process is the *vortex motion* in water drops falling through the NMR magnet
- Due to the short residence time of the drop in the receiver coil, *single-point acquisition* is necessary for all points in k and q_v space
- Drops from different dripping experiments show different velocity profiles
- A *water drop* covered with a surfactant does not show internal motion
- A 2D velocity vector field is composed of two data sets, each providing one of the two in-plane velocity components
- If k or q_v space is traced in only one or two dimensions without line or slice selection, a projection is acquired in real space or in velocity space (see *projection – cross-section theorem* at the end of this chapter), i.e. the signal is integrated over the hidden space or velocity components

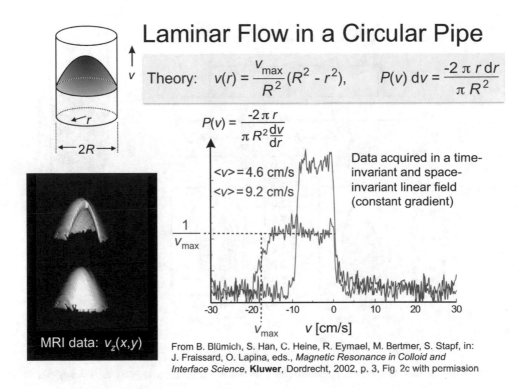

Laminar Flow in a Circular Pipe

Theory: $v(r) = \dfrac{v_{max}}{R^2}(R^2 - r^2),$ $\quad P(v)\,dv = \dfrac{-2\pi r\,dr}{\pi R^2}$

$$P(v) = \dfrac{-2\pi r}{\pi R^2 \dfrac{dv}{dr}}$$

$<v> = 4.6$ cm/s

$<v> = 9.2$ cm/s

$\dfrac{1}{v_{max}}$

Data acquired in a time-invariant and space-invariant linear field (constant gradient)

-30   -20   -10   0   10   20   30

v_{max} v [cm/s]

MRI data: $v_z(x,y)$

From B. Blümich, S. Han, C. Heine, R. Eymael, M. Bertmer, S. Stapf, in: J. Fraissard, O. Lapina, eds., *Magnetic Resonance in Colloid and Interface Science*, **Kluwer**, Dordrecht, 2002, p. 3, Fig. 2c with permission

Probability Densities of Velocity

- *Probability densities* are called *distributions* in short
- A *distribution of velocities* is the Fourier transform of the NMR signal as a function of $\boldsymbol{q}_v$
- In the simplest experiment one component of $\boldsymbol{q}_v$ is varied and the distribution is plotted against *displacement R* in a given time Δ
- This notation is also applied to diffusive motion. In the NMR community, the corresponding *distribution of displacements* is called the *propagator*
- The distribution of velocities is most simply measured with a pair of anti-phase field-gradient pulses, and the amplitude of the associated echo is recorded as a function of the gradient amplitude
- In principle, the experiment can be conducted also with time-invariant field gradients by varying the echo time t_E in an echo experiment
- For *laminar flow through a circular pipe* the *velocity profile* is parabolic
- The velocity distribution is obtained by equating the probability $P(v)\,dv$ of finding a velocity component between v and $v + dv$ with the relative area of the annulus at radius r with width dr in which this velocity component is found
- The distribution is constant for all velocities between 0 near the tube wall and v_{max} in the center, and zero elsewhere. It has the shape of a hat
- *Velocity distributions* are very sensitive against slight imperfections in the experimental set-up. They may provide better fingerprints of a flow process than *velocity images* and are faster to measure

2D Velocity Distributions

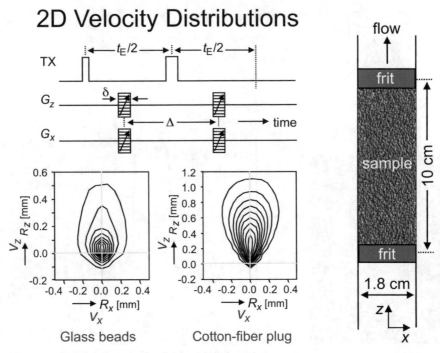

From B. Blümich, *NMR Imaging of Materials*, **Clarendon Press**, Oxford, 2000, Fig. 5.4.7 by permission of Oxford University Press

Velocity-Vector Distributions

- 1D *velocity distributions* are *projections* of 3D velocity distributions obtained by integration over the unresolved velocity components
- *2D velocity distributions* provide more detailed information than 1D distributions
- They are obtained by measuring the transverse magnetization as a function of two components of the wave vector q_v and subsequent 2D Fourier transformation
- Flow through circular pipes filled with glass beads or cotton fibers can readily be distinguished by the associated 2D distributions of radial and axial velocities
- For the cotton fibers, strong radial dispersion is observed at high axial flow
- For the glass beads, considerable axial backflow is observed at zero radial flow
- The observed velocities are *finite difference approximations* of velocities corresponding to displacements R experienced during the encoding time Δ
- For field gradient pulses with durations δ no longer short compared to the characteristic times of motion, the *slow-motion approximation* fails and the finite difference interpretation can no longer be applied

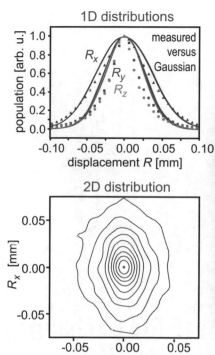

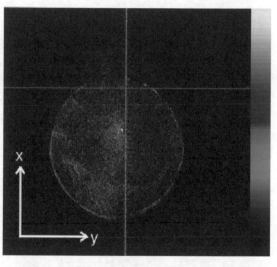

Oriented Ice from Freezing Salt Water

From M. Menzel, S.-I. Han, S. Stapf, B. Blümich, **J. Magn. Reson. 143** (2000) 376, Figs. 2, 3, 4 with permission

Diffusion in Anisotropic Media

- *Translational diffusion* leads to *incoherent displacements* of molecules as opposed to *coherent displacement* by flow
- To probe incoherent motion, displacements R_x, R_y, R_z are measured in given time intervals Δ with PFG NMR using the same pulse sequences as for flow encoding in a first approximation
- For free diffusion, the *distribution of displacements* (*propagator*) has a Gaussian form, and the mean diffusion length $<R^2>^{1/2}$ scales with the square root of the diffusion time Δ
- For free 3D diffusion, one obtains $<R^2> = <(r_2 - r_1)^2> = 6\,D\,\Delta$, where D is the diffusion coefficient. For 1D diffusion on obtains $<R^2> = 2\,D\,\Delta$
- For *restricted diffusion* in the narrow pores of rocks and heterogeneous catalysts, the confining pore walls limit the diffusion length
- For short diffusion times, a *Gaussian distribution* is observed. For long diffusion times, the confinements lead to deviations from a Gaussian distribution
- Macroscopic anisotropy of porous media, such as oriented biological tissue and ice formed from salt water can be identified by comparing the 1D distributions of displacements in different space directions at long diffusion times
- Another method consist of analyzing *2D distributions of displacements* for deviations from circular symmetry

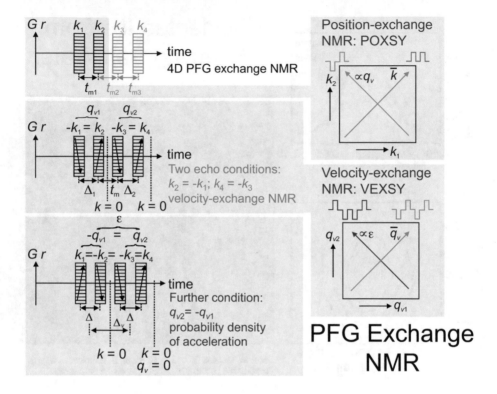

Position-Exchange NMR

- *Flow* or *displacements* in given times are typically measured with short, anti-phase gradient-pulse pairs. The second, positive pulse marks final position and the first, negative pulse initial position
- For measurements of displacements, both gradients pulses are locked to equal magnitude in each amplitude step, i.e. $k_2 = -k_1$ at all times
- Stepping both gradient pulses independently leads to a 2D experiment
- The Fourier transform of the acquired data is the joint probability density of finding a magnetization component at a particular initial position and a time Δ later at a particular final position
- On the principal diagonal, the average of the initial and final positions is identified. On the secondary diagonal, the difference between final and initial positions, i.e. displacement or velocity, is identified
- The experiment is called *position-exchange spectroscopy* (*POXSY*) in analogy to *frequency exchange-spectroscopy* (EXSY) in NMR spectroscopy
- A 4D exchange experiment results with four gradient pulses
- Conditions on the gradient variations, such as the formation of a *gradient echo* for detection, reduce the dimensionality of the experiment
- The conditions $k_2 = -k_1$ and $k_4 = -k_3$ imposed on the 4D POXSY sequence lead to *velocity-exchange spectroscopy* (*VEXSY*)
- Here, *average velocity* is identified along the principal diagonal and velocity difference or *acceleration* along the secondary diagonal
- The additional condition $k_4 = -k_3 = -k_2 = k_1$ leads to a 1D experiment by which the *distribution of accelerations* can be measured

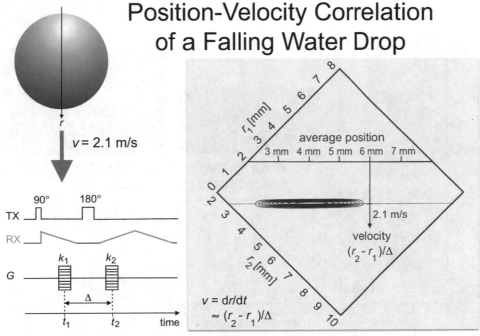

From S.-I. Han, S. Stapf, B. Blümich in: J. Fraissard, O. Lapina, eds., *Magnetic Resonance in Colloid and Interface Science*, **Kluwer**, Dordrecht, 2002, p. 327, Fig. 5 with permission

Demonstration of Position-Exchange NMR

- An instructive example of a *position-exchange experiment* is that performed on a falling *water drop*
- Initial and final positions are marked in the Fourier domain by the wave numbers k_1 and k_2
- The maximum of a Hahn or a stimulated echo with one gradient pulse in each free evolution period is recorded for all values of k_1 and k_2
- The 2D Fourier transform of the experimental data set is the position-exchange map
- The principal diagonal identifies average position and the secondary diagonal position difference
- Along the principal diagonal, the projection of the drop onto the gradient direction is observed
- Along the secondary diagonal, the drop displacement during the encoding time Δ corresponding to the *velocity* of the falling drop is observed

Cross Filtration by VEXSY

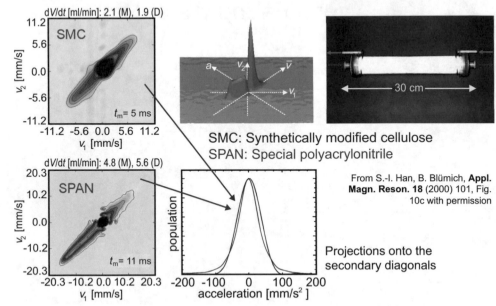

SMC: Synthetically modified cellulose
SPAN: Special polyacrylonitrile

From S.-I. Han, B. Blümich, **Appl. Magn. Reson. 18** (2000) 101, Fig. 10c with permission

Projections onto the secondary diagonals

From B. Blümich, S. Han, C. Heine, R. Eymael, M. Bertmer, S. Stapf, in: J. Fraissard, O. Lapina, eds., *Magnetic Resonance in Colloid and Interface Science*, **Kluwer**, Dordrecht, 2002, p. 3, Fig. 6 with permission

Velocity-Exchange NMR

- Similar to the *position-exchange NMR* experiment, initial and final velocities can be encoded in terms of q_{v1} and q_{v2} along the axes of a 2D data matrix, leading to the *velocity-exchange NMR* experiment (*VEXSY*)
- Velocity-exchange NMR has been used to study *cross-filtration* in hollow-fiber filtration modules which are used in hemodialysis as artificial kidneys
- Water was passed inside and outside the hollow fibers in counter flow
- Water molecules crossing the membrane must change their direction and lead to off-diagonal signal in a velocity-exchange map
- On the diagonal, the distribution of *average velocity* is observed
- For negative velocities one finds the hat function corresponding to *laminar flow* within the round membrane fibers. For positive velocities, the distribution maps the interstitial flow and depends on the packing of the fibers
- The projection along the principal diagonal and onto the secondary diagonal of the velocity-exchange map eliminates average velocity from the data set, and the remaining variable is velocity difference or *acceleration*
- For two different membrane materials (SMC, SPAN) and operating conditions, the velocity exchange spectra are different, and so are the projections onto the secondary diagonals, i.e. the *distributions of accelerations*
- SPAN shows signal at high accelerations corresponding to more efficient interactions of the molecules with the membrane walls than the SMC

Projection - Cross-Section Theorem

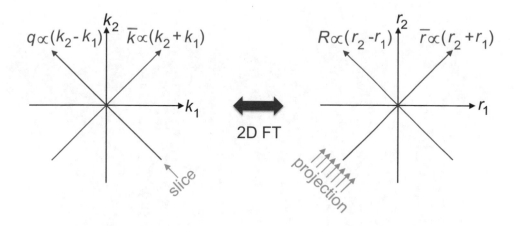

$$q \propto (k_2 - k_1) \quad \overline{k} \propto (k_2 + k_1)$$

$$R \propto (r_2 - r_1) \quad \overline{r} \propto (r_2 + r_1)$$

2D FT

Projections in Multi-Dimensional Fourier NMR

- In *multi-dimensional Fourier NMR* the data are acquired in the space of one set of variables (t, $\mathbf{k}$, $\mathbf{q}_v$) and subsequently Fourier transformed into the space of the set of *Fourier-conjugate variables* (ω, $\mathbf{r}$, $\mathbf{v}$) appropriate for data interpretation
- A *slice* in one space corresponds to a *projection* in the Fourier-conjugate space
- For example, a 1D slice in (k_1,k_2) space along the secondary diagonal corresponds to a projection in Fourier-conjugate space along the principal diagonal so that the data are projected onto the secondary diagonal
- The term "projection" means "integration" of the multivariable function so that the number of variables is reduced and the variable in the direction of the projection is the variable of the integration
- This relationship can readily be derived from the expression for the *multi-dimensional Fourier transformation*
- It is known as the *projection – cross-section theorem*
- The projection – cross-section theorem is helpful in relating 1D distributions to 2D distributions and designing fast experiments with reduced dimensionality

5. Relaxometry and Laplace NMR

Distributions of relaxation times
Distributions of diffusion coefficients
Diffusive diffraction
Compact magnets
Well-logging NMR
2D Laplace NMR
Depth profiling
Applications to materials characterization

© Springer Nature Switzerland AG 2019
B. Blümich, *Essential NMR*,
https://doi.org/10.1007/978-3-030-10704-8_5

Acquiring Distributions of Relaxation Times

From B. Blümich in: R.A. Mayers, ed., *Enc. Anal. Chem.*, **John Wiley**, Chichester, 2016, a9458, Fig. 13 with permission

NMR Relaxometry

- *Relaxation times* and *diffusion coefficients* can be measured in inhomogeneous magnetic fields by analyzing signal decay and build-up in the *time domain*
- *Inhomogeneous fields* are inexpensive to generate with *permanent magnets*
- The fields of permanent magnets are low (up to 2 T) compared to those of most superconducting NMR magnets (up to 24 T)
- NMR relaxometry with compact low-field instruments was commercialized by Bruker in 1973 for food analysis and not much later by Oxford Instruments
- It is used to test the physical properties of materials, because relaxation times and self-diffusion coefficients report on the rotational and translational molecular motion and not the chemical structure
- Longitudinal relaxation is measured with the *inversion recovery sequence* and the *saturation recovery sequence*
- Transverse relaxation in liquids is measured with the *Hahn echo* and the *CPMG sequence*. In solids it is measured with the *solid echo* and Ostroff-Waugh (*OW4*) sequences, which use 90° instead of 180° refocusing pulses
- In simple liquids, T_1 and T_2 relaxation is mono-exponential
- In multi-component mixtures such as emulsions and fluids in porous media, the relaxation signal can be approximated by a sum of exponential functions
- This can be inverted to a relaxation-time distribution by operations similar to *inverse Laplace transformation*
- This is why *time-domain NMR* is often also called *Laplace NMR*

Fluids in Porous Media

- The signal amplitude is proportional to *porosity* at full fluid saturation
- The longitudinal relaxation rate $1/T_1$ and the transverse relaxation rate $1/T_2$ are enhanced by *wall relaxation*
- The *transverse relaxation rate* is affected by self diffusion in applied and in *internal gradients*
- The molecular *self-diffusion* length is confined by the pore diameter

Relaxation of fluids in in porous media:

$$\frac{1}{T_2} = \frac{1}{T_{2,\text{bulk}}} + \rho_2 \frac{S}{V} + \frac{D\,(\gamma\,G\,t_E)^2}{12}$$

$$\frac{1}{T_1} = \frac{1}{T_{1,\text{bulk}}} + \rho_1 \frac{S}{V}$$

S: surface area
V: pore volume
ρ: surface relaxivity
t_E: echo time
D: apparent diffusion coefficient

Restricted diffusion in fluid-filled pores:

$$\frac{D}{D_0} = 1 - \frac{4}{9\sqrt{\pi}} \frac{S}{V} \sqrt{D_0\,\Delta} + O(\Delta)$$

D: apparent diffusion coefficient
D_0: bulk diffusion coefficient
Δ: diffusion time

Relaxation and Diffusion

- In fluid-filled *porous media*, relaxation and diffusion reveal essential material properties of fluid and matrix, which can readily be measured by ^{1}H NMR
- The signal amplitude is proportional to the amount of fluid in the sensitive volume, i.e. to the *fluid saturation* for partially saturated porous media and the *porosity* for fully saturated porous media
- The pore walls confine the diffusion paths and enhance the relaxation rates of the fluid molecules in the pores
- In the *fast diffusion limit* the relaxation rate is proportional to the *surface-to-volume ratio*, i.e. the relaxation time reports *pore size*
- Contrary to the longitudinal relaxation rate, the transverse relaxation rate is affected by translational diffusion in inhomogeneous fields with a gradient G
- The *field gradient* can arise from the inhomogeneity of the applied field and from *susceptibility differences* due to the heterogeneity of the porous medium
- Information about the *pore size* is also obtained from the restrictions in molecular self-diffusion imposed by the pore walls
- NMR relaxation and self-diffusion signals are inverted to *distributions of relaxation rates* and *distributions of apparent diffusion coefficients* identifying in a porous medium, for example, *bound fluid* and *producible fluid*
- Such distributions from fluid-saturated porous media are interpreted in terms of *pore-size distributions*, which are evaluated for *porosity*, *permeability*, and *fluid typing* in *well-logging NMR*

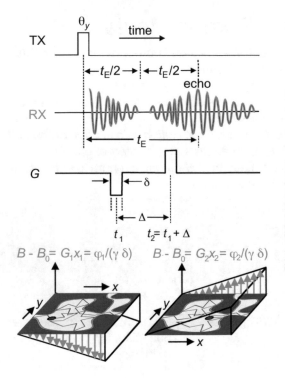

Gradient Echo

Measuring displacement
from diffusion or flow

NMR with pulsed
gradient fields:
PFG NMR
(pulsed field-
gradient NMR)

Measuring Diffusion

- Like velocity, *diffusion* is measured in terms of *displacement* $R = r_2 - r_1$ within a given time window, where r is the position of the molecule
- Position is encoded by the phase $\varphi = \gamma\, G\, r\, \delta = k\, r$ of the transverse magnetization evolving for a time δ in an *inhomogeneous field*, which is preferably linear with a constant gradient G. Here k is the position wave number
- Displacement is measured by the phase $\Delta\varphi = k_1\, r_1 + k_2\, r_2 = k\, (r_2 - r_1) = q\, R$ accumulated with two gradient pulses at different times having the same wave-number magnitude k but opposite signs, where q is the wave number for displacement
- In such an *anti-phase gradient-pulse pair* the position wave number $k = k_1 + k_2$ vanishes after the second gradient pulse so that position is not encoded in the signal phase and the gradient-pulse pair is said to generate a *gradient echo*
- If all spins move coherently in a pulsed linear gradient-field, all acquire the same phase difference, which appears as a *phase shift of the echo*
- If the spins diffuse randomly, positive and negative phase shifts cancel on average, and the echo amplitude is attenuated according to $M_{xy}(t) = M_{xy}(0)$ $\exp\{-t_E/T_2\} \times \exp\{-q^2 D(\Delta - \delta/3)\}$, where D is the *self-diffusion coefficient* and Δ the *diffusion time*
- The transverse magnetization $M_{xy}(t) = M_0 \int P(R) \exp\{-i\, q\, R\}\, dR$ of an ensemble of spins diffusing different distances R during the diffusion time without relaxation is determined by the Fourier transform of the *probability density* $P(R)$ *of displacements*, which is known as the *propagator* in NMR

Pulse Sequences for Diffusion NMR

PFG NMR with a Hahn echo

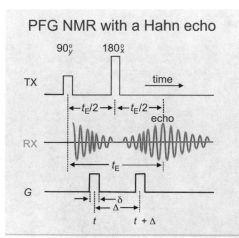

CPMG sequence in a time-invariant gradient field $B - B_0 = G\,r$ and the effective gradient G_{eff}

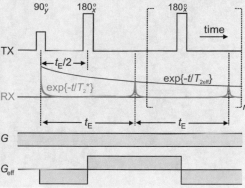

Distortions of the magnetic induction field $\mathbf{B}$ by susceptibility differences:
$\Delta\chi = \chi_{pore} - \chi_{matrix}$

Estimation of the *internal gradient* of a pore:
$G_{int} \approx \Delta\chi\, H\,/\,r_{pore}$
H: vacuum magnetic field

Time-Invariant Gradient Fields

- If the field gradient results from the sensor or *magnetic susceptibility differences* in the object it is time-invariant and an echo can be generated by rf pulses
- In the maximum of the Hahn echo and the stimulated echo, the dephasing of magnetization components from magnetic fields that are inhomogeneous across the sample is refocused unless the spins are moving
- The refocusing rf pulse of the Hahn echo changes the sign of all *precession phases* $\varphi = k\,r$ accumulated before
- This is equivalent to a sign change of all position-encoding gradients active before, and one can write the time variation in terms of the *effective gradient* G_{eff}, which gives the same result but without refocusing rf pulses
- For example, identical gradient pulses before and after the refocusing pulse in a Hahn echo to encode *diffusion* and *coherent flow* are equivalent to an echo with the effective gradient
- In a CPMG train applied in a linear field $B(r) - B_0 = G\,r$ all odd echoes refocus dephasing from different positions, and all even echoes in addition also refocus dephasing from coherent displacement
- The decay of the *CPMG echo envelope* due to *relaxation* and *diffusion* is given by $M_{xy}(t) = M_{xy}(0)\,\exp\{-n\,t_E\,/\,T_2\}\,\exp\{-q^2\,D\,n\,t_E\,/\,3\}$
- In *stray-field NMR*, the rf field is inhomogeneous and all rf pulses are selective, so that the flip angles vary across the sample
- Then the CPMG echoes are mixed Hahn and stimulated echoes, and T_2 becomes an *effective relaxation time* T_{2eff}, which is a mixture of T_1 and T_2

Effects of Displacement

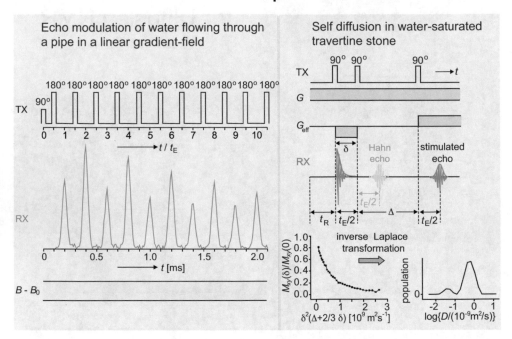

Displacing Spins

- The paths of spins being displaced during a time shorter than T_1 can be probed by different NMR methods
- In a *stimulated echo* sequence, the echo amplitude is attenuated from displacement by *diffusion D* in a constant gradient and by relaxation following
  $$M_{xy}(t) = \tfrac{1}{2} M_{xy}(0) \times \exp\{-q^2 D (\Delta + t_E/3)\} \times \exp\{-\Delta/T_1\} \times \exp\{-t_E/T_2\}$$
- Other than diffusion, *laminar flow* through a pipe is coherent leading to a phase shift in addition to the attenuation of the echo from diffusion and relaxation
- As $q = \gamma G t_E/2$, the logarithm the CPMG *echo attenuation* depends linearly on t_E^3 for free diffusion, and D can be determined from the slope of $\ln\{M_{xy}\}$
- To improve the sensitivity of diffusion measurements, the diffusion-attenuated echo may be refocused several times with echo time t_{E2} and the echo sum be detected for a systematic variation of the first echo time t_{E1}
- For sufficiently short echo times t_{E2} in the CPMG *detection period*, signal loss in the *CPMG echo train* due to diffusion is suppressed
- Then, without the summation, the scheme gives a 2D data set of Hahn-echo attenuation from diffusion versus T_2 relaxation decays
- The T_2 relaxation can be neglected in the *diffusion filter* for strong gradients, and a D-T_2 correlation map is obtained by *2D inverse Laplace transformation*
- In porous media, diffusion is restricted, so that the signal attenuation by diffusion is lowered, and that from relaxation is enhanced due to *wall relaxation*

Waves in Space

$M_{xy}(\boldsymbol{r}) = M_{xy}(\boldsymbol{0}) \exp\{i\,\boldsymbol{k}\,\boldsymbol{r}\}$
$= M_{xy}(\boldsymbol{0})\,[\cos\{\boldsymbol{k}\,\boldsymbol{r}\} + i\,\sin\{\boldsymbol{k}\,\boldsymbol{r}\}]$

$\boldsymbol{r}$: space vector
$\boldsymbol{k}$: wave vector
$\boldsymbol{k}\,\boldsymbol{r} = \varphi$: phase

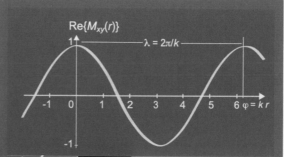

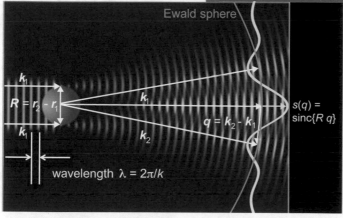

Diffraction

The interference patterns of light, X-rays, neutrons, or electrons report the Fourier transform of the distribution of scattering centers

Distributions of Velocities and Displacements

- According to the phase table in chapter 4, the magnetization component moving with constant velocity v through a linear gradient-field is given by
  $M_{xy}(t) = M_{xy}(0)\,\exp\{-t/T_2\}\,\exp\{-i\varphi_{11}\}$
- Here $\varphi_{11}(t) = \gamma_0 \int^t G(t')\,t'\,dt'\,v = \gamma_0 \int^t G(t')\,t'\,dt'\,R/\Delta$, where R is the distance travelled during time Δ and $G(t)$ is the *effective gradient*
- For coherent flow $\varphi_{11}(t) = q_v(t)\,v$. In the case of a diffusing magnetization component, the notation of Callaghan applies, and $\varphi_{11}(t) = q(t)\,R$ with $k = q$
- For many magnetization components the transverse magnetization is the integral over all velocities v or displacements R,
  $M_{xy}(t) = (1/V) \int M_{xy}(v)\,\exp\{-i\,q_v\,v\}\,dv = (1/V) \int M_{xy}(R)\,\exp\{-i\,q\,R\}\,dR$
- $M_{xy}(v)/V$ is the *velocity distribution* and $M_{xy}(R)/V$ the *displacement distribution*
- For constant magnetization density, $M_{xy}(R)/(M_0\,V) = P(R)$ is the probability density of displacements, also known as the *propagator*
- For diffusion, R depends on the diffusion time Δ and so does the propagator
- For free diffusion, the propagator is a *Gauss function*, which broadens with increasing diffusion time, $P(R,\Delta) = (4\pi\,D\,\Delta)^{-3/2}\,\exp\{-R^2/(4\,D\,\Delta)\}$, the variance $<R^2> = 2\,D\,\Delta$ of which defines the *mean diffusion length* $<R^2>^{1/2} = (2\,D\,\Delta)^{1/2}$
- For *restricted diffusion* in identical pores, the propagator is the *auto-correlation function of the pore shape* in the fast diffusion limit, and *diffraction* peaks are observed in the NMR signal

Diffraction

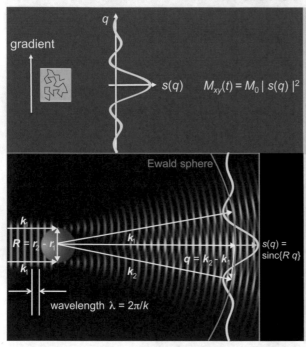

Diffusive diffraction of nuclear magnetization waves

In the fast diffusion limit the transverse magnetization is the magnitude square of the Fourier transform of the pore-shape function: $M_{xy}(t) = M_0 \, |\, s(q)\, |^2$

Diffraction of conventional waves

The interference pattern is described by the magnitude square of the Fourier transform $s(q)$ of the distribution of scattering centers: $I(q) = I_0 \, |\, s(q)\, |^2$

In the first diffraction maximum: $q = 2\pi/R$

Diffusive Diffraction

- The *propagator* $P(R)$ can be written as integral over all initial positions r_1 of the product of the probability density $S(r_1)$ of finding a spin at r_1 and the conditional probability density $S(r_1, r_2, \Delta)$ of finding it at $r_2 = r_1 + R$ some time Δ later,

$$P(R) = \int S^*(r_1)\, S(r_1,\, r_1 + R,\, \Delta)\, dr_1$$

- At long diffusion time Δ, $S(r_1,\, r_1 + R,\, \Delta) = S(r_2)$. Then, with $q\, R = k\,(r_2 - r_1)$,

$M_{xy}(q) = M_0 \int P(R) \exp\{-i\, q\, R\}\, dR$, and rewriting $P(R)$ one obtains

$$M_{xy}(q) = M_0 \int\!\!\int S^*(r_1)\, S(r_2)\, dr_1 \exp\{-i\, k\,(r_2 - r_1)\}\, d(r_2 - r_1)$$

$$= M_0 \int\!\!\int S^*(r_1)\, S(r_2)\, \exp\{-i\, k\, r_2\}\, \exp\{i\, k\, r_1\}\, d(r_2 - r_1)\, dr_1$$

$$= M_0 \int S^*(r_1)\, \exp\{i\, k\, r_1\}\, dr_1 \int S(r_2)\, \exp\{-i\, k\, r_2\}\, dr_2$$

$$= M_0\, s^*(k)\, s(k) = M_0\, |\, s(k)\, |^2 = M_0\, |\, s(q)\, |^2$$

- This states, that the transverse magnetization measured in a diffusion experiment reports the magnitude square of the Fourier transform of the probability density $S(r)$ of finding spins at position r
- For fluid-filled pores, $S(r)$ is the normalized projection of the pore shape along the gradient direction. It is commonly called the *pore-shape function*
- In the limit of fast diffusion or long Δ, the transverse magnetization is proportional to the magnitude square of the Fourier transform of the pore-shape function

Compact Magnets for Laplace NMR

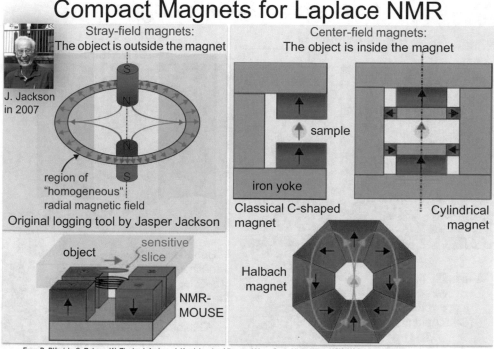

From B. Blümich, C. Rehorn, W. Zia, in: J. Anders, J. Korvink, eds., *Micro and Nano Scale NMR*, **Wiley-VCH**, Weinheim, 2018, Fig. 1.7 with permission

Permanent Magnets for NMR Relaxometry

- The first NMR experiments were done with *electromagnets* and later on also with permanent magnets at ^{1}H frequencies from 20 MHz up to 90 MHz
- Permanent magnets were hard to shim and drift with temperature so that compact permanent magnets were mostly used for relaxation and diffusion measurements in the past
- There are two basic types of magnets for NMR, stray-field magnets and center-field magnets
- *Center-field magnets* accommodate the sample inside, so that the sample size is limited by the diameter of the magnet opening
- The classical geometry is C-shaped, while mirrored C-shapes and cylinder shapes provide better field homogeneity due to higher symmetry
- *Halbach magnets* are particularly efficient and ideally are without stray field
- *Stray-field magnets* produce fields that enter the object from one side
- They were first used for *well logging* to be inserted into the hole of an oil well and to acquire signal from the fluids in the borehole wall
- Compact stray-field instruments were subsequently developed mostly for moisture measurements in buildings and porous media
- The *NMR-MOUSE* (Mobile Universal Surface Explorer) is a small stray-field sensor, which collects the signal from a flat sensitive slice outside the magnet
- It can be used to measure depth profiles into the surface of arbitrarily large objects by changing the distance between sensor and object

Concepts of NMR Logging

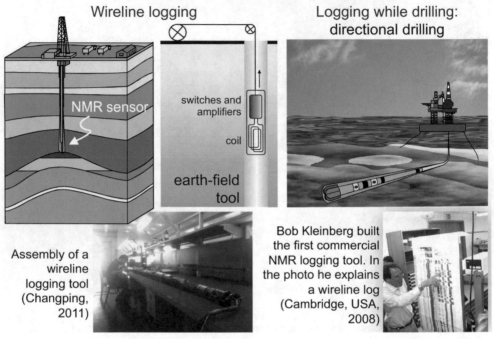

Wireline logging

Logging while drilling:
directional drilling

NMR sensor

switches and
amplifiers

coil

earth-field
tool

Assembly of a
wireline
logging tool
(Changping,
2011)

Bob Kleinberg built
the first commercial
NMR logging tool. In
the photo he explains
a wireline log
(Cambridge, USA,
2008)

NMR Well Logging

- Early on NMR was of interest for *well logging*. Russell Varian had the idea in the early 1950's, and oil companies like Chevron, Mobil, Texaco, Borg-Warner and Schlumberger started to study fluids in porous rock
- The first generation nuclear magnetic logging tools operated in the earth's magnetic field. Later on J. Jackson introduced permanent magnets
- Since the magnet is inside the sample and not the sample inside the magnet, this type of stray-field NMR is also called *inside-out NMR*
- Due to the inhomogeneous stray field and field-gradients inside the rock pores induced by susceptibility differences, the NMR signal is acquired with CPMG-like multi-echo sequences
- Since 1995 NMR logging tools are operated by service companies. *Halliburton* calls their tool MRIL (Magnetic Resonance Imaging Log) and *Schlumberger* calls it CMR (Combinable Magnetic Resonance tool)
- The tools of the first and the second generations were operated in *wireline mode* after drilling and measured the NMR information while being pulled up through the hole
- The third generation tools are designed for *logging while drilling*. They are part of the drill string and suitable for *directional drilling*
- Some logging-while-drilling tools explore the original magnet geometry proposed by Jasper Jackson

Sensor Designs

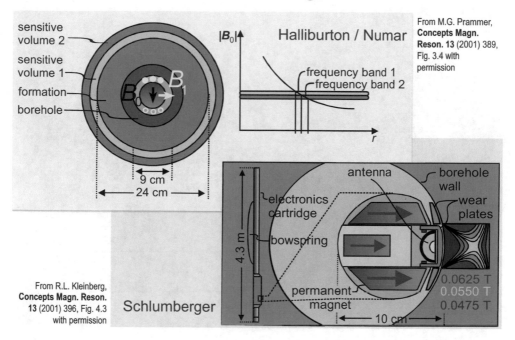

sensitive volume 2

sensitive volume 1

formation

borehole

$|B_0|$

Halliburton / Numar

frequency band 1
frequency band 2

r

From M.G. Prammer, **Concepts Magn. Reson. 13** (2001) 389, Fig. 3.4 with permission

9 cm
24 cm

antenna
borehole wall
electronics cartridge
wear plates
bowspring
4.3 m

permanent magnet

0.0625 T
0.0550 T
0.0475 T

10 cm

From R.L. Kleinberg, **Concepts Magn. Reson. 13** (2001) 396, Fig. 4.3 with permission

Schlumberger

NMR Logging Tools

- The *earth-field tool* of the first generation used a strong DC current through the coil to magnetize the protons and then used the same coil for detection
- The *deadtime* was 30 ms, and the fluid molecules exposed to fast *wall relaxation* could not be measured
- Only *free fluid* could be measured, but not *porosity*. Oil, water, and bore-hole fluid could not be distinguished
- By fitting the sensors with magnets, several problems could be solved:
  - The NMR frequency increased from 2 kHz to about 1 MHz so that the deadtime was shortened and the sensitivity increased
  - The signal dependence on the sensor orientation in the earth's field is avoided
  - The sensitive volume can be localized inside the borehole wall away from the borehole fluid
- The *Jackson tool* has a sensitive volume in the shape of an annulus, which is narrow in the axial direction. Measurements with such a sensor moving along the bore-hole are difficult as the spins find little time to be polarized
- The *Halliburton/Numar tool* has transverse dipolar B_0 and B_1 fields, which are largely perpendicular to each other at all locations. The field magnitude at a given radius is constant along the circumference. The radial gradient helps to select signal from different shells by tuning to different frequencies
- The *Schlumberger tool* provides an axially extended sweet spot in one angular segment at some distance away from the sensor

NMR of Fluids in Porous Media

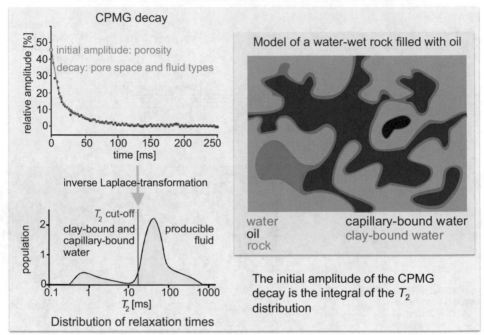

CPMG decay

initial amplitude: porosity
decay: pore space and fluid types

inverse Laplace-transformation

T_2 cut-off
clay-bound and capillary-bound water — producible fluid

Distribution of relaxation times

Model of a water-wet rock filled with oil

water | capillary-bound water
oil | clay-bound water
rock

The initial amplitude of the CPMG decay is the integral of the T_2 distribution

Formation Evaluation by NMR

- Modern *NMR logging tools* acquire data from more than 5 cm deep into the formation of the bore-hole wall
- Typically the transverse magnetization decay from fluid-filled pores is measured as the envelope of a Carr-Purcell-Meiboom-Gill (*CPMG*) *echo train*
- Its amplitude provides mineralogy-independent *porosity* $\Phi = V_{pores}/V_{rock}$
- The *distribution of relaxation times* T_2 is obtained from the echo-train decay envelope by *inverse Laplace transformation*
- The area under the distribution curve is the amplitude of the CPMG signal
- A model-bound analysis of the relaxation time distribution yields
  - *irreducible water saturation* from *clay-bound* and *capillary-bound water*
  - amount *of producible fluid*
  - *permeability estimates*
  - *hydrocarbon type*
  - *oil-viscosity estimates*
- Porosity is estimated traditionally from density values determined by *gamma-ray scattering* or *neutron scattering*, applying formation-dependent corrections
- NMR porosity is formation independent unless the formation is magnetic
- Advanced *multi-dimensional Laplace methods* provide more detailed information, e.g. they can separate signals from oil and water and characterize *pore connectivity*

NMR Porosity and Relaxation in Pores

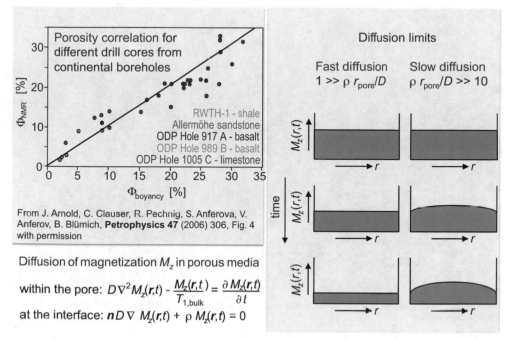

Porosity correlation for different drill cores from continental boreholes

Φ_{NMR} [%]

$\Phi_{boyancy}$ [%]

RWTH-1 - shale
Allermöhe sandstone
ODP Hole 917 A - basalt
ODP Hole 989 B - basalt
ODP Hole 1005 C - limestone

From J. Arnold, C. Clauser, R. Pechnig, S. Anferova, V. Anferov, B. Blümich, **Petrophysics 47** (2006) 306, Fig. 4 with permission

Diffusion limits

Fast diffusion
$1 >> \rho\, r_{pore}/D$

Slow diffusion
$\rho\, r_{pore}/D >> 10$

Diffusion of magnetization M_z in porous media

within the pore: $D\nabla^2 M_z(r,t) - \dfrac{M_z(r,t)}{T_{1,bulk}} = \dfrac{\partial M_z(r,t)}{\partial t}$

at the interface: $\mathbf{n}D\nabla\, M_z(r,t) + \rho\, M_z(r,t) = 0$

Porosity and Surface Relaxation

- The *porosity* Φ is directly determined from the signal amplitude of the fluid in the rock normalized to the signal amplitude of bulk water and corrected for the *hydrogen index* HI, which corresponds to the relative spin density of the fluid
- The fluid molecules diffuse inside the pores of the rock matrix
- To reduce the signal loss within the *deadtime* of the instrument, low measuring fields are advantageous because lower fields induce smaller *field inhomogeneity* at interfaces resulting from *magnetic susceptibility differences* so that signal attenuation from *diffusion in internal gradients* of the pores is low
- NMR porosity scales well with porosity determined by *buoyancy*
- The *longitudinal magnetization* $M_z(r,t)$ follows the *diffusion equation* with the *bulk relaxation time* $T_{1,bulk}$ within a pore, and in the direction along the normal $\mathbf{n}$ of the wall it relaxes with the *surface relaxivity* ρ
- A normal mode analysis leads to the slow and the fast diffusion limits
- In the *fast diffusion limit*, the relaxation rate scales with the *surface-to-volume ratio* S/V, where $S/V = 3/r_{pore}$ for spherical pores, and the transverse magnetization decay is mono-exponential
- The fast diffusion limit is assumed to be valid in most applications of NMR to *well logging*
- In the *slow diffusion regime*, the magnetization decay is multi-exponential for a single type of pore and depends on the *pore geometry*

Relaxation-Time Distributions and Pore Size

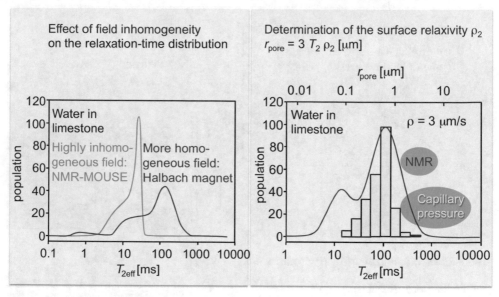

From J. Arnold, C. Clauser, R. Pechnig, S. Anferova, V. Anferov, B. Blümich, **Petrophysics 47** (2006) 306, Figs. 5. 6 with permission

Surface Relaxation

- The *surface relaxivity* ρ links *pore size* and *relaxation time*
- It is determined by matching the *relaxation-time distribution* from NMR with the *pore-throat distribution* determined e.g. by differential capillary pressure measurement such as *mercury intrusion porosimetry*
- The pore-throat distribution scales with the derivative of the *capillary pressure* (CP) curve versus fluid saturation
- The assumptions are that the pore radii are proportional to the throat radii and that all pores have the same surface relaxivity
- By comparing T_1 and T_2 relaxation-time distributions, one finds, that most rocks exhibit T_1/T_2 ratios between 1 and 2.5 with a most probable value of 1.65 at 2 MHz. That value becomes larger at higher measuring fields
- Deviations may arise at short relaxation times from contributions to T_2 by diffusion in internal gradients if the echo time is not chosen short enough
- The shapes of the NMR relaxation-time distributions depend only weakly on pressure and temperature
- For immobile molecules $T_{2\text{eff}}$ derived from a multi-echo decay in a strongly inhomogeneous magnetic field is prolonged by scrambling of longitudinal and transverse magnetization components from resonance-offset related flip-angle distributions
- For mobile molecules $T_{2\text{eff}}$ is shortened by diffusion in field gradients

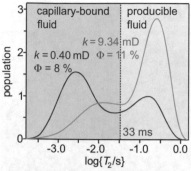

Producing Water from Porous Media

T_2 distributions of Aller-moehe sandstones: The bound water decreases with increasing permeability k and macro porosity Φ of the rock

From S. Anferova, V. Anferov, J. Arnold, E. Talnishnikh, M.A. Voda, K. Kupferschläger, P. Blümler, C. Clauser, B. Blümich, **Magn. Reson. Imag. 25** (2007) 474, Fig. 6 with permission

Estimating permeability

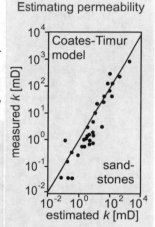

From C.-C. Huang, *Estimation of Rock Properties by NMR Relaxation Methods*, **Thesis**, Rice University, Houston, 1997, Fig. 6.5.?.? with permission

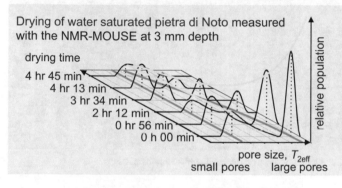

Drying of water saturated pietra di Noto measured with the NMR-MOUSE at 3 mm depth

Fluid Production from Porous Media

- When drying a water-wet rock, the large pores lose water first and the small pores later, i.e. the irreducible water remains in the small pores
- A *relaxation-time cut-off* serves to estimate *bound water* from *producible fluid*
- The T_2 cut-off for water is 33 ms in sandstones and 92 ms in carbonates
- For *crude oil* with a broad distribution, the relaxation times of bound water and producible oil may overlap, but the fractions may be distinguished by diffusion measurements
- The flow velocity v of a fluid with dynamic viscosity η through a porous medium in a pressure gradient $\Delta p/\Delta x$ is determined by the *permeability* k, $v = k/\eta \, \Delta p/\Delta x$. It is measured in Darcy, where 1 Darcy = 1 D = 0.987×10^{-12} m^2
- The permeability $k = C \, d^2$ is poportional to the square of the effective pore diameter d, where C is a constant relating to the flow paths
- It is estimated from the relaxation-time distribution following models of the form $k = a \, \Phi^4 \, T_{2,k}^2$, where, depending on the model, $T_{2,k}$ is an average relaxation time derived from all or parts of the relaxation-time distribution
- The *Kenyon model* fits best for 100% brine saturated rock, the *Coates-Timur model* applies to hydrocarbons, and the *Chang model* to vuggy carbonates
- The Chang model has been improved to accommodate a large range of permeabilities by introducing a tortuosity-dependent weight
- The *tortuosity* $\tau = D_0/D_\infty$ is the ratio of the bulk self-diffusion coefficient to the apparent self-diffusion coefficient at long diffusion time $\Delta \to \infty$

Example of a Gas-Well Log

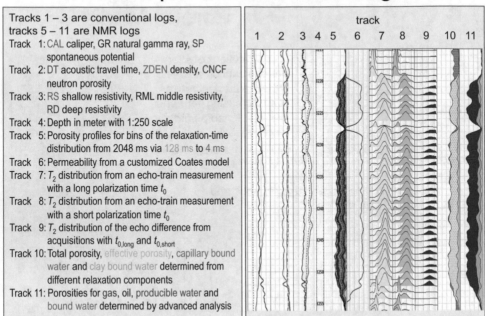

Tracks 1 – 3 are conventional logs, tracks 5 – 11 are NMR logs

Track 1: CAL caliper, GR natural gamma ray, SP spontaneous potential

Track 2: DT acoustic travel time, ZDEN density, CNCF neutron porosity

Track 3: RS shallow resistivity, RML middle resistivity, RD deep resistivity

Track 4: Depth in meter with 1:250 scale

Track 5: Porosity profiles for bins of the relaxation-time distribution from 2048 ms via 128 ms to 4 ms

Track 6: Permeability from a customized Coates model

Track 7: T_2 distribution from an echo-train measurement with a long polarization time t_0

Track 8: T_2 distribution from an echo-train measurement with a short polarization time t_0

Track 9: T_2 distribution of the echo difference from acquisitions with $t_{0,long}$ and $t_{0,short}$

Track 10: Total porosity, effective porosity, capillary bound water and clay bound water determined from different relaxation components

Track 11: Porosities for gas, oil, producible water and bound water determined by advanced analysis

Courtesy of L. Xiao, China University of Petroleum, Changping, China

Elementary Well Logging

- Elementary NMR *well logging* is based on the analysis of *CPMG decays* and associated *distributions of relaxation times* from fluid saturated pores
- The amplitude of the total signal estimates the porosity
- The amplitude of the slowly relaxing part estimates the amount of *producible fluid* (free fluid index FFI) corresponding to the integral of the relaxation time distribution to the right of the relaxation cut-off
- The amplitude of the rapidly relaxing components estimates the amount of capillary and clay-bound water also called the bulk volume irreducible BVI corresponding to the integral of the relaxation time distribution to the left of the relaxation cut-off
- *Permeability* is estimated from the porosity following different models
- Information about the hydrocarbon type is obtained from T_1 *weights* introduced by different *polarization times* t_0 and from *diffusion weights* introduced by different *echo times* t_E when the signal is acquired in the presence of a field gradient
- All NMR data are interpreted in the context of other logging data from measurements of *electrical resistivity*, *gamma-ray scattering*, *neutron scattering*, etc.
- Detailed measurements by *2D Laplace NMR* and other more advanced methods are performed in *logging-while-drilling* applications and on core samples in the laboratory
- Well logging data are validated by data from laboratory studies

Hydrogen Index and Relaxation of Dead Oils

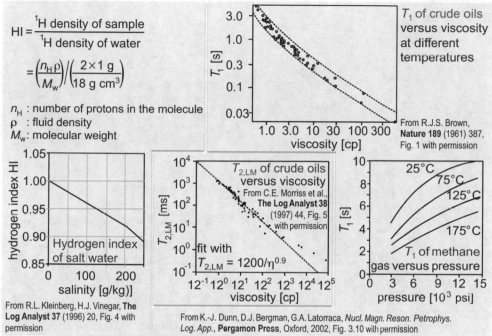

$$HI = \frac{{}^1H \text{ density of sample}}{{}^1H \text{ density of water}}$$

$$= \left(\frac{n_H \rho}{M_w}\right) \bigg/ \left(\frac{2 \times 1\,g}{18\,g\,cm^3}\right)$$

n_H : number of protons in the molecule
ρ : fluid density
M_w: molecular weight

From R.L. Kleinberg, H.J. Vinegar, **The Log Analyst 37** (1996) 20, Fig. 4 with permission

From R.J.S. Brown, **Nature 189** (1961) 387, Fig. 1 with permission

From C.E. Morriss et al., **The Log Analyst 38** (1997) 44, Fig. 5 with permission

From K.-J. Dunn, D.J. Bergman, G.A. Latorraca, *Nucl. Magn. Reson. Petrophys. Log. App.*, **Pergamon Press**, Oxford, 2002, Fig. 3.10 with permission

NMR Properties of Fluids in Porous Media

- The NMR signals from fluids in porous media relate to the bulk fluids and the pore space. The signal decay is affected by *bulk relaxation*, *wall relaxation*, and by *diffusion* across regions with different magnetic field strengths
- The *signal amplitude* scales with the total number of protons within the sensitive volume. At full fluid saturation it reports formation *porosity*
- Natural fluids saturating rock are water of variable salinity, gas, and oil
- The *water salinity* ranges from zero to full saturation with mostly NaCl
- The most important gas in well logging is *methane*
- When the oil is free of gas, it is called *dead oil*, otherwise it is called *live oil*
- Live oils exist at elevated pressures and temperatures below the bubble point. Their relaxation times depend on the the gas/oil ratio
- *Synthetic oils* are used in *drilling mud* and can mix with reservoir fluids
- The amount of a fluid is quantified from the NMR signal amplitude via the proton density relative to that of water. This quantity is the *hydrogen index HI*
- Since the beginning of the relaxation curve is hidden within the *dead time* of the instrument, *HI* determined by NMR is an *apparent hydrogen index*
- Most rocks are water wet, so that the detected water signal exhibits relaxation times shortened by *wall relaxation*
- Because the relaxation time of bulk water is long (1 to 3 s), the bulk-water signal can be suppressed with short repetition times

Bulk Diffusion of Water, Gas, and Oil

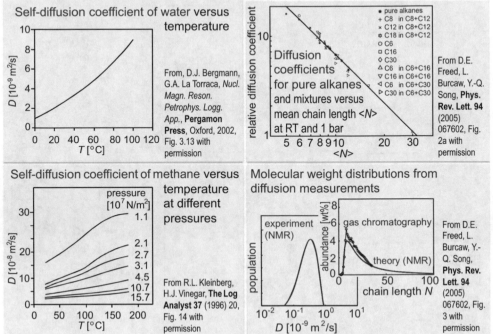

Self-diffusion coefficient of water versus temperature

From, D.J. Bergmann, G.A. La Torraca, *Nucl. Magn. Reson. Petrophys. Logg. App.*, **Pergamon Press**, Oxford, 2002, Fig. 3.13 with permission

Diffusion coefficients for pure alkanes and mixtures versus mean chain length $<N>$ at RT and 1 bar

- pure alkanes
- + C8 in C8+C12
- × C12 in C8+C12
- ✻ C18 in C8+C12
- ○ C6
- □ C16
- ◇ C30
- △ C6 in C6+C16
- ▽ C16 in C6+C16
- ◁ C6 in C6+C30
- ▷ C30 in C6+C30

From D.E. Freed, L. Burcaw, Y.-Q. Song, **Phys. Rev. Lett. 94** (2005) 067602, Fig. 2a with permission

Self-diffusion coefficient of methane versus temperature at different pressures

pressure [10^7 N/m^2]
1.1
2.1
2.7
3.1
4.5
10.7
15.7

From R.L. Kleinberg, H.J. Vinegar, **The Log Analyst 37** (1996) 20, Fig. 14 with permission

Molecular weight distributions from diffusion measurements

experiment (NMR)
gas chromatography
theory (NMR)

From D.E. Freed, L. Burcaw, Y.-Q. Song, **Phys. Rev. Lett. 94** (2005) 067602, Fig. 3 with permission

Bulk Relaxation and Diffusion

- In water-wet rock the oil is not in direct contact with the pore wall, so that T_1 and T_2 of oil in the pores are the same as in bulk
- For pure compounds both relaxations are mono-exponential, but *crude oils* consist of many components and exhibit distributions of relaxation times
- From the T_2 distribution $x(T_2)$, the *logarithmic mean relaxation time* $T_{2,LM}$ is calculated as $\log\{T_{2,LM}/s\} = [\Sigma_i \, x_i \log\{T_{2i}/s\}]/[\Sigma_i \, x_i]$. It scales with the *viscosity*
- For *dead oils* $T_{2,LM}$ scales linearly with the ratio of viscosity/temperature
- Unlike T_1 of liquids, T_1 of gases is shortened by the *spin rotation interaction*, i.e. the coupling between the proton spins and the field produced by the rotation of the molecule, so that T_1 of gases increases with increasing pressure
- In inhomogeneous fields the magnetization decays by T_2 and diffusion
- The *diffusion of water* (D = 2.23 10^{-9} m²/s at RT) is insensitive to pressure but sensitive to temperature
- The *diffusion of methane* ($D \approx 10^{-3}$ m²/s at RT) is much faster than that of water and increases with increasing temperature and decreasing density
- For a melt of linear polymers, the diffusion coefficient $D_i = N_i^{-\nu} A <N>^{-\beta}$ of component i relates to the average degree $<N>$ of polymerization through a power law with degrees of polymerization N_i and exponents ν, β
- The *molecular weight distribution* of the melt is mapped by the distribution of diffusion coefficients, i.e. the chemical composition of crude oil can be mapped by the distribution of a physical NMR parameter

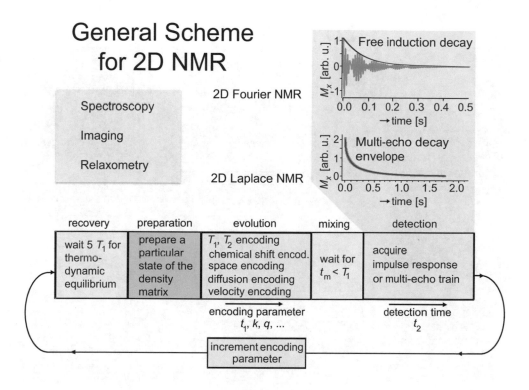

General Two-Dimensional NMR

- Two-dimensional NMR was introduced for *high-resolution spectrosocpy* and subsequently extended to 3D and *multi-dimensional NMR spectroscopy*
- With the availability of a fast *2D inverse Laplace transformation algorithm*, 2D Laplace NMR became popular
- Similar to the *1D inverse Laplace transformation*, also the 2D inverse Laplace transformation is unstable in the presence of noise and needs to be regularized, so that the transforms are not unique but, depending on the regularization algorithm, they are probable solutions
- The scheme for *2D Laplace NMR* is the same as that for *2D Fourier NMR* including MRI, except that instead of a free-induction decay (FID), a multi-echo decay envelope is recorded during the detection period t_2
- It's amplitude is modulated by the events during the preceding *mixing period*, *evolution period*, *preparation period* and *recovery period*
- The most popular 2D Laplace experiments are the T_1-T_2 correlation experiment, the T_2-T_2 exchange experiment and the D-T_2 correlation experiment
- *Diffusion-ordered spectroscopy* (*DOSY*) can be understood as a mixed 2D Laplace-Fourier NMR experiment, which correlates distributions of diffusion coefficients with distributions of chemical shifts, i.e. with the NMR spectrum. It is used for chemical identification of components in complex solutions

2D Laplace NMR

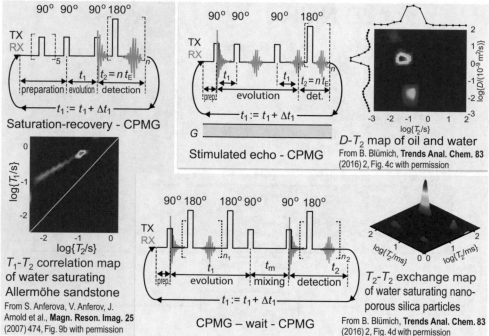

Saturation-recovery - CPMG

Stimulated echo - CPMG

T_1-T_2 correlation map
of water saturating
Allermöhe sandstone
From S. Anferova, V. Anferov, J.
Arnold et al., **Magn. Reson. Imag. 25**
(2007) 474, Fig. 9b with permission

CPMG – wait - CPMG

D-T_2 map of oil and water
From B. Blümich, **Trends Anal. Chem. 83**
(2016) 2, Fig. 4c with permission

T_2-T_2 exchange map
of water saturating nano-
porous silica particles
From B. Blümich, **Trends Anal. Chem. 83**
(2016) 2, Fig. 4d with permission

Two-Dimensional Laplace NMR

- *Two-dimensional Laplace NMR* concerns correlation maps of *distributions of relaxation times* and *diffusion coefficients*. It can be executed on fluids in *porous media* exhibiting *internal gradients* and even with stray-field instruments
- The signal is detected with a multi-echo sequence giving the T_2 distribution
- With T_1-T_2 *correlation NMR*, T_1 is encoded in the evolution period
- The T_1-T_2 correlation map of water in Allermöhe sandstone reports the impact of diffusion in internal gradients G on the transverse signal decay: Due to diffusive attenuation, the T_1/T_2 ratio varies with pore size $1/(S/V)$
- With D-T_2 *correlation NMR*, D is encoded in the evolution period
- While the relaxation times of oil and water in porous rock overlap on the T_2 scale, the two types of fluid can be separated along the diffusion axis D
- D-T_2 correlation NMR is good for quantifying different fluids such as oil and water in porous media based on differences in diffusion coefficients and T_2
- T_2 -T_2 *exchange NMR* probes diffusion without application of field gradients by molecules diffusing between different relaxation sites
- T_2 is encoded in both, the evolution period and the detection period, which are separated by a *mixing time* t_m
- The exchange map reveals connectivity of pores by diffusion-mediated magnetization transfer between pores similar to *2D NOESY spectroscopy*, which reveals distances between chemical sites by relaxation-mediated magnetization transfer between different chemical groups in molecules

Diffusion-Ordered Spectroscopy

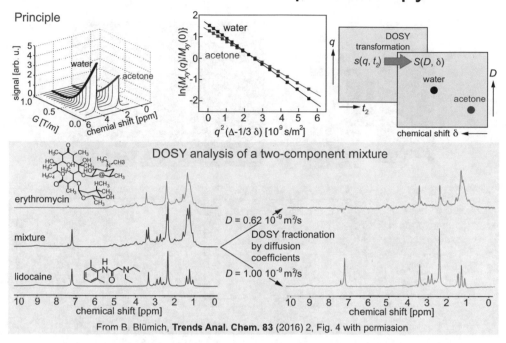

From B. Blümich, **Trends Anal. Chem. 83** (2016) 2, Fig. 4 with permission

2D Fourier-Laplace NMR

- Mixed 2D Fourier-Laplace methods correlate the NMR spectrum with a distribution of relaxation times or diffusion coefficents, where the spectral information is detected directly during t_2
- The most powerful method is *diffusion-ordered spectroscopy* (*DOSY*)
- It separates a crowded spectrum of a multi-component liquid into a set of subspectra of components with similar diffusion coefficients
- In the pulse sequence, the FID is diffusion encoded in the evolution periods of a stimulated echo with two anti-phase field-gradient pulses of amplitude G and duration δ. With the diffusion time Δ the signal of one component follows the decay $M_{xy}(t) = \frac{1}{2} M_{xy}(0) \times \exp\{-q^2 D (\Delta - \delta/3)\} \times \exp\{-\Delta/T_1\} \times \exp\{-t_E/T_2\}$
- Instead of successive Fourier and Laplace transformations a dedicated DOSY transformation provides better inversion stability
- Slices through diffusion peaks of the DOSY map $S(D,\delta)$ parallel to the chemical shift axis δ produce separate spectra for molecules with distinct diffusion coefficients D
- DOSY can be understood as a non-destructive form of *spectral editing* by fractionation through diffusion

Depth Profiling with the NMR-MOUSE

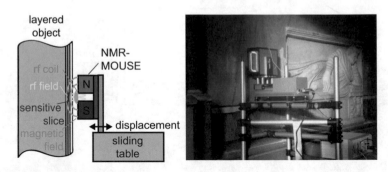

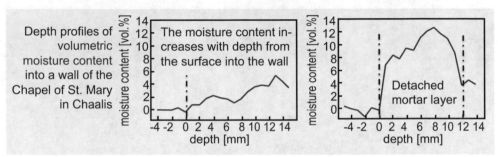

Depth profiles of volumetric moisture content into a wall of the Chapel of St. Mary in Chaalis

The moisture content increases with depth from the surface into the wall

Detached mortar layer

Depth Profiling

- The *NMR-MOUSE* (MObile Universal Surface Explorer) is a *stray-field NMR relaxometer* with a thin, flat slice, which defines the sensitive volume at a fixed distance up to 25 mm away from the surface of the device
- The spins in the sensitive slice produce the NMR signal
- Typically CPMG-type multi-echo trains are measured at different distances of the sensor surface from the object surface
- The NMR signal $M_{xy}(t) = \Sigma_i\, M_{xy,i}(0) \exp\{-t/T_{2\text{eff},i}\}$ is analyzed for relative component amplitudes $x_i = M_{xy,i}(0)/M_{xy}(0)$ and relaxation times $T_{2\text{eff},i}$ to define amplitudes of profiles as a function of depth into the object (1D images)
- The initial magnetizations $M_{xy,i}(0)$ can be sensitized to diffusion or longitudinal relaxation by means for suitable pulse sequences preceding signal detection
- For fluids in porous media, the initial amplitude of the echo envelope gives *volumetric fluid content* when normalized to the amplitude of the bulk fluid
- The normalized integral of the *CPMG echo envelope* defines a spin-density weighted *average relaxation time*, $_0\!\int^\infty M_{xy}(t)\, \mathrm{d}t\, /\, M_{xy}(0) = \Sigma_i\, x_i\, T_{2\text{eff},i} = <T_{2\text{eff}}>$
- The empirical *weight function* $w = (t_2\text{-}t_1)/t_1\, _0\!\int^{t_1} M_{xy}(t)\, \mathrm{d}t\, /\, _{t_1}\!\int^{t_2} M_{xy}(t)\, \mathrm{d}t$ provides good contrast as depth-profile amplitude at favorable signal-to-noise ratio

Droplet Size and Solid-Fat Content

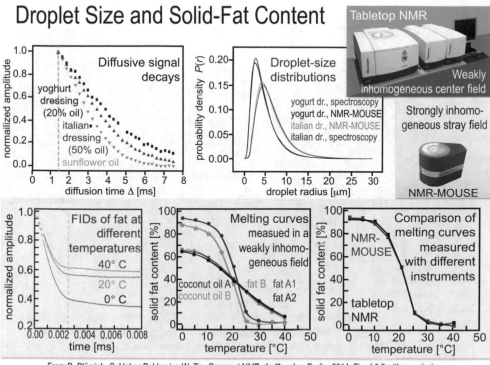

From B. Blümich, S. Haber-Pohlmeier, W. Zia, *Compact NMR*, **de Gruyter**, Berlin, 2014, Fig. 4.3.2 with permission

Emulsions and Suspensions

- *Emulsions* are mixtures of immiscible fluids, where one phase is dispersed as droplets in another continuous phase. Examples are mixtures of oil and water like milk, salad dressing, and skin cream
- In strong field gradients the signal from the freely diffusing molecules decays fast, so that for oil-in-water emulsions, the oil signal survives the water signal
- After 1 to 2 ms diffusion time Δ in a stimulated echo sequence, the water signal in salad dressing has decayed in the stray-field gradient of the order of 20 T/m from the NMR-MOUSE, and the decay of the surviving signal is dominated by the *restricted diffusion* in the more viscous oil droplets
- The decay observed as a function of the diffusion time Δ is typically fitted with a decay modeled for a logarithmic Gaussian *droplet-size distribution*
- *Suspensions* are solid particles suspended in a liquid. They are found in processed food and fat
- The FID from the solid decays fast, that from the liquid much slower
- NMR relaxometry is an official method to determine the *solid fat content*, which specifies the ratio of protons from solid fat to those from total fat
- With the *indirect method*, the amount of liquid fat is given by the room-temperature amplitude of fat at 25 μs deadtime, and the amount of total fat by the signal amplitude at elevated temperature of 60° C where all fat is molten

Skin and Plants

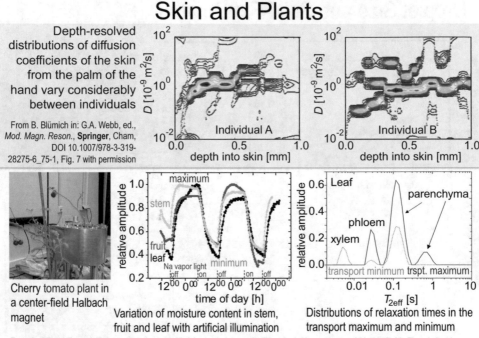

Depth-resolved distributions of diffusion coefficients of the skin from the palm of the hand vary considerably between individuals

From B. Blümich in: G.A. Webb, ed., *Mod. Magn. Reson.*, **Springer**, Cham, DOI 10.1007/978-3-319-28275-6_75-1, Fig. 7 with permission

Cherry tomato plant in a center-field Halbach magnet

Variation of moisture content in stem, fruit and leaf with artificial illumination

Distributions of relaxation times in the transport maximum and minimum

From D. Oligschläger, C. Rehorn, S. Lehmkuhl, M. Adams, A. Adams, B. Blümich, **J. Magn. Reson. 278** (2017) 80, Figs. 4, 5 with permission

Biological Tissues

- *Biological tissues* are assemblies of cells, which are found in animals and plants
- The most important application of NMR is the study of *human tissue* by *MRI*
- The *contrast* in medical MRI is largely determined by spin density, relaxation times and diffusion coefficients, which are all accessible with Laplace NMR
- The *NMR-MOUSE* has been modified to work with a 2 mm thick sensitive slice with a homogeneous gradient for frequency encoding of depth
- This version has been employed to measure depth-resolved distributions of diffusion coefficients through human *skin*. They bear the characteristic signature of the individual
- With a simple permanent center-field magnet, small *plants* can be studied and the water in the pant tissue be characterized
- The amount of water in different parts of the plant is given by the NMR signal amplitude
- The water transport is high when it is dark and the light stress is off
- *Distributions of relaxation times* measured in the water-concentration extrema during the dark and bright periods identify different locations of the water in the plant tissue

Elastomers are
cross-linked
rubber molecules

Relaxometry of Cross-Linked Rubber

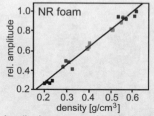

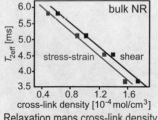

bulk NR

stress-strain   shear

$T_{2\text{eff}}$ [ms]

cross-link density [10^{-4} mol/cm^3]

Relaxation maps cross-link density

NR foam

rel. amplitude

density [g/cm^3]

Amplitude maps gravimetric density

Transverse magnetization decays map cross-link density via relaxation time and gravimetric density via signal amplitude. For strained elastomers the decays depend on the orientation of the strain direction in the magnetic field

From J. Kolz, J. Martins, K. Kremer, T. Mang, B. Blümich, **KGK 60** (2009) 179, Figs. 7, 8 with permission

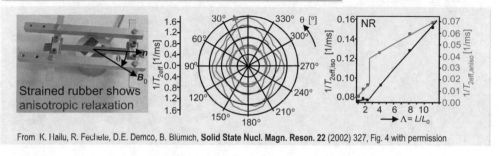

Strained rubber shows anisotropic relaxation

$1/T_{2\text{eff}}$ [1/ms]

θ [°]

$1/T_{2\text{eff,iso}}$ [1/ms]

$1/T_{2\text{eff,aniso}}$ [1/ms]

NR

$\Lambda = L/L_0$

From K. Hailu, R. Fechete, D.E. Demco, B. Blümich, **Solid State Nucl. Magn. Reson. 22** (2002) 327, Fig. 4 with permission

Rubber and Elastomers

- *Rubber* is a melt of entangled macromolecules
- *Elastomers* are chemically crosslinked macromolecules in the molten state
- Sulfur is the most common crosslinking agent
- The chemical crosslinking process is called *vulcanization*
- *Technical elastomers* are statistical products from macromolecules, fillers like carbon black, and processing aids, made by mixing, diffusion, and reaction
- In rubber and elastomers the ^{1}H *relaxation times* are determined by the *chain stiffness* and the *residual dipole-dipole coupling* among ^{1}H nuclei
- A fraction of the dipole-dipole interaction remains unaveraged due to the motional anisotropy of chain segments between entanglements and crosslinks
- At 60 to 80 K above the *glass temperature* T_g the motion is fast, and $T_{2\text{eff}}$ is proportional to the *crosslink density* determined by measuring strain and shear
- Closer to T_g the chain stiffness also determines the relaxation times
- The signal amplitude of foamed rubber scales with the *foam density*
- For strained elastomers the relaxation rate depends on the orientation angle
- It is the sum of an isotropic and an anisotropic component. The latter can be approximated by a *Gaussian distribution* of *second Legendre polynomials*
- For NR both relaxation rates change like first-order and second-order phase transitions at the point where *strain-induced crystallization* sets in

Curing, Aging, and Processing

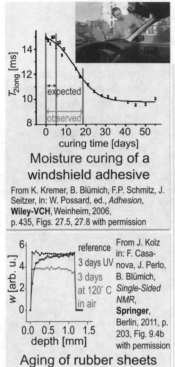

Moisture curing of a windshield adhesive

From K. Kremer, B. Blümich, F.P. Schmitz, J. Seitzer, in: W. Possard, ed., *Adhesion*, **Wiley-VCH**, Weinheim, 2006, p. 435, Figs. 27.5, 27.8 with permission

reference
3 days UV
3 days at 120° C in air

From J. Kolz in: F. Casanova, J. Perlo, B. Blümich, *Single-Sided NMR*, **Springer**, Berlin, 2011, p. 203, Fig. 9.4b with permission

Aging of rubber sheets

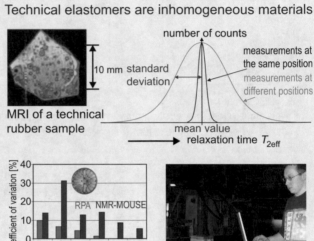

Technical elastomers are inhomogeneous materials

MRI of a technical rubber sample

From B. Blümich, A. Buda, K. Kremer, **GAK 6** (2006) 290, Fig. 2 with permission

Properties of Elastomers

- T_{2eff} changes with the *curing time*, e.g. for a moisture-curing poly(urethane) adhesive used for mounting windshields in cars
- Changes in T_{2eff} can be observed well beyond the nominal curing time
- *Aging* of rubber in the presence of oxygen is mostly a combination of chain scission, cross-linking, and oxidation of the polymer
- In many cases the rubber hardens upon *aging*
- *UV aging* affects only the surface of the rubber, while *thermo-oxidative aging* also affects deeper layers
- Technical rubber products are inhomogeneous due to their statistical nature
- Depending on the miscibility of compounds and the processing conditions, some compounds may agglomerate as revealed by *NMR imaging*
- The reproducibility of T_{2eff} in measurements at one point is better than 1 %, but may be worse than 10 % for measurements at different points due to the statistical nature of the material and the small volume of the NMR-MOUSE
- The *NMR-MOUSE* collects signal from a volume element much smaller than the sample size for the rubber process analyzer (RPA), giving rise to a larger coefficient of variation
- For *quality control of rubber products* several measurements need to be conducted at equivalent spots of the product. The mean fit parameters characterize the *average material properties* and the standard deviation the *material heterogeneity*

Non-Destructive Tire Testing

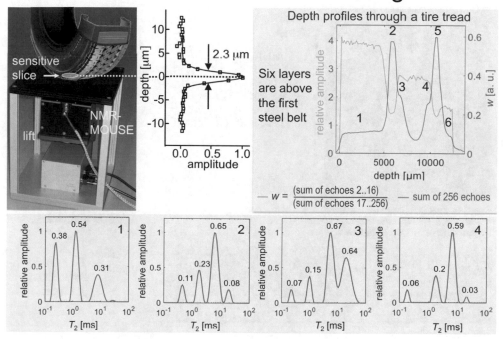

— $w = \dfrac{\text{(sum of echoes 2..16)}}{\text{(sum of echoes 17..256)}}$ — sum of 256 echoes

Quality Control of Elastomer Products

- At room temperature, the relaxation times of rubber and elastomers strongly depend on temperature
- For *quality control* of *rubber products*, experimental values of $T_{2\text{eff}}$ need to be extrapolated to a reference temperature and correlated with material properties by means of calibration curves
- *Calibration curves* are determined with small samples in physical testing laboratories by measuring swelling, rheology, and dynamic-mechanical relaxation
- By means of calibration curves NMR relaxation times report parameters like the *glass temperature*, the *elastic modulus*, and the *crosslink density* non-destructively in selected spots from production intermediates and the final product
- A standard application of the NMR-MOUSE is for *tire testing*
- The depth profile of a tire reveals all layers from the tire tread, including the tread base, the steel-belt overlay, and the belt coat up to the first steel belt
- At each depth, the amplitude of the depth profile is calculated in different ways from the *CPMG decay*
- The *distributions of relaxation times* obtained by Laplace transformation of the CPMG decays fingerprint the properties of the material at the given depth

Morphology of Semi-Crystalline Polymers

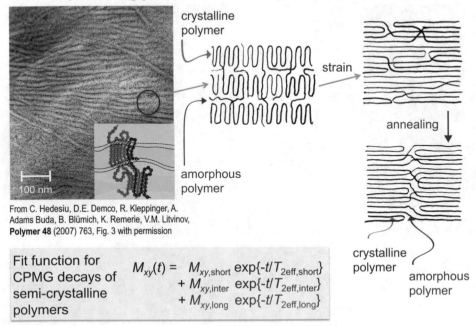

crystalline polymer

strain

annealing

amorphous polymer

crystalline polymer

amorphous polymer

From C. Hedesiu, D.E. Demco, R. Kleppinger, A. Adams Buda, B. Blümich, K. Remerie, V.M. Litvinov, **Polymer 48** (2007) 763, Fig. 3 with permission

Fit function for CPMG decays of semi-crystalline polymers

$$M_{xy}(t) = M_{xy,short}\, \exp\{-t/T_{2eff,short}\}$$
$$+ M_{xy,inter}\, \exp\{-t/T_{2eff,inter}\}$$
$$+ M_{xy,long}\, \exp\{-t/T_{2eff,long}\}$$

Semi-Crystalline Polymers

- *Semi-crystalline polymers* are solids from macromolecules with mobile amorphous and rigid crystalline domains. Sometimes a rigid amorphous interface can be identified by transverse relaxation measurements
- In the *crystalline domains*, the chains are ordered and packed densely, the degrees of freedom for molecular motion are restricted, and T_2 is short
- In the *amorphous domains*, the chains are disordered, the packing density is lower, the motional degrees of freedom are less restricted, and T_2 is longer
- Semi-crystalline polymers like poly(ethylene), poly(propylene), and Nylon, solidify from the melt in lamellar stacks of crystalline and amorphous domains
- Upon drawing, the macromolecular chains realign
- Upon annealing, a fibrillar structure forms with alternating layers of chain-folded lamellae and amorphous domains arranged perpendicular to the drawing direction
- Depending on the temperature history and the shear fields applied during processing, different overall *crystallinity*, different size distribution of the crystallites, and different order in the amorphous domains are generated
- A crude fit to the *transverse relaxation decay* of such polymers is a *tri-exponential function*, where the rapidly, intermediate and slowly relaxing components are assigned to the crystalline, the interfacial and the amorphous domains, respectively

The Impact of Temperature on PE

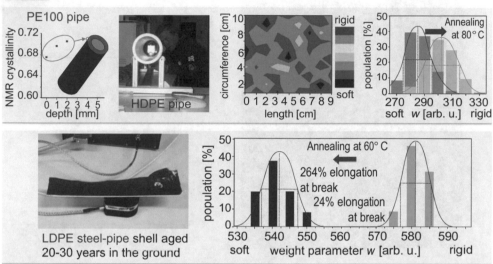

From B. Blümich, A. Adams-Buda, M. Baias, **GWF Gas Erdgas 148** (2007) 95, Figs. 2, 3, 4 with permission

Heterogeneity and Annealing of PE

- *Polymer products* are inhomogeneous for a number of reasons
- They are made from macromolecules with distributions in molecular porperties such as molecular weight, branching, and copolymer statistics
- Polymer processing often involves cooling from the melt. Rapid cooling at the surface leads to lower crystallinity than slow cooling inside the bulk
- In semi-crystalline polymers also the domain sizes are distributed
- The transverse relaxation decays contain signal contributions from crystalline, interfacial and amorphous domains
- The rapidly decaying signal from the crystalline domains is partially lost in the deadtime of the NMR instrument
- A higher value of the *weight parameter w* extracted from the CPMG decay indicates more rigid material or higher crystallinity
- In a melt-extruded HDPE pipe the *crystallinity* increases from the surface to the interior due to the temperature gradient during cooling
- The small sensitive volume of the NMR-MOUSE maps fluctuations in crystallinity. Their distribution shifts to higher crystallinity upon annealing for 24 h at a temperature of 80° C well below the melting temperature of 136° C
- An old LDPE steel-pipe shell has become brittle due to chemical corrosion. The elastic properties could be restored by annealing for 24 h at a temperature of 60° C

Deformation of PE Pipes

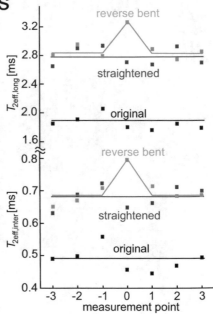

From A. Adams, M. Adams, B. Blümich, J.H. Kocks, O. Hilgert, S. Zimmermann, **3R Int. 49** (2010) 216, Figs. 8, 20 with permission

Mechanical Properties of Semi-Crystalline Polymer Materials

- The *mechanical properties* of semi-crystalline polymer materials are usually explained in terms of the degree of *crystallinity*
- This is so, because crystallinity can easily be determined by *X-ray diffraction* and thermo-analytical methods, whereas NMR is a less common method
- In contrast to X-ray diffraction, NMR relaxation directly measures the signal from the *amorphous domains*
- At small loads in the linear regime of the *stress-strain curve*, deformations are reversible and largely affect only the amorphous but not the crystalline domains
- This is evidenced by the changes in relaxation time $T_{2\text{eff}}$ of a slightly curved, carbon-black filled section of *water pipe* cut from a pipe delivered on a large spool
- Upon bending the section straight, the $T_{2\text{eff}}$ values of the amorphous and interface domains change over the whole length of the pipe section but not those of the crystalline domains
- Upon bending the section in reverse over a pin, the relaxation times of the amorphous and interface domains change only locally at the site of the pin

Physical and Chemical Aging of LDPE

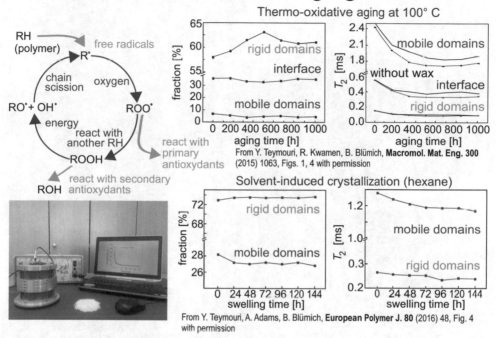

From Y. Teymouri, R. Kwamen, B. Blümich, **Macromol. Mat. Eng. 300** (2015) 1063, Figs. 1, 4 with permission

From Y. Teymouri, A. Adams, B. Blümich, **European Polymer J. 80** (2016) 48, Fig. 4 with permission

Aging of Semi-Crystalline Polymers

- *Aging* of *semi-crystalline polymers* involves at least two competitive processes: *Physical aging* (*crystallization*) and *chemical aging* (*corrosion*)
- A third process can be the loss of low molecular weight compounds by evaporation and leaching through *solvent exposure*
- During *thermo-oxidative aging* the material is exposed to elevated temperature in air
- At short aging times the material predominantly crystallizes so that the fraction of crystalline domains increases, and the mobility, i.e. T_2, of the amorphous domains decreases
- At longer aging times, chemical *degradation* dominates along with chain scission causing the rigid fraction to decrease and the mobility in the amorphous domains to increase
- *Wax* is generated in most polymerization processes as evidenced by the number fraction in the *Schulz-Flory distribution* of molecular weights
- The presence of wax accelerates the physical aging process
- Solvents are taken up essentially only by the amorphous domains in semi-crystalline polymers
- The chain mobility gained by the amorphous chains upon swelling leads to *solvent-induced crystallization* reminiscent of *temperature-induced crystallization*. This process may be impacted by the leaching of wax

Paintings

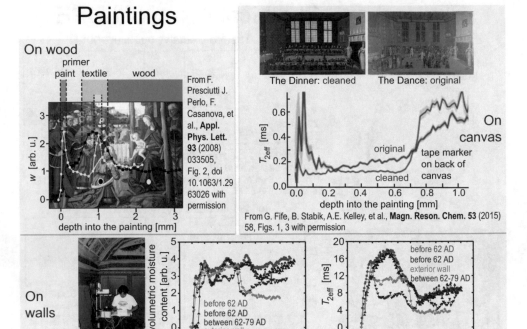

On wood

primer
paint textile   wood

w [arb. u.]

depth into the painting [mm]

From F. Presciutti J. Perlo, F. Casanova, et al., **Appl. Phys. Lett. 93** (2008) 033505, Fig. 2, doi 10.1063/1.2963026 with permission

The Dinner: cleaned   The Dance: original

$T_{2\text{eff}}$ [ms]

original

tape marker on back of canvas

cleaned

On canvas

depth into the painting [mm]

From G. Fife, B. Stabik, A.E. Kelley, et al., **Magn. Reson. Chem. 53** (2015) 58, Figs. 1, 3 with permission

On walls

volumetric moisture content [arb. u.]

before 62 AD
before 62 AD between 62-79 AD exterior wall

depth [mm]

$T_{2\text{eff}}$ [ms]

before 62 AD
before 62 AD exterior wall
between 62-79 AD

depth [mm]

From A. Haber, B. Blümich, D. Souvorova, E. Del Federico, **Anal. Bioanal. Chem. 401** (2011) 1441, Fig. 4 with permission

Cultural Heritage

- Non-destructive testing by stray-field relaxation and diffusion NMR finds more and more applications for studying objects of value to *cultural heritage*
- A proton density higher than expected in depth profiles often indicates the presence of a *conservation agent* like wax in mortar or varnish in bones
- The *stratigraphy of paintings* reveals the thickness of planar layers with an accuracy of 10 μm and material properties from relaxation times and diffusion constants
- For example, in the 1473 AD painting on wood 'Adoration of the Magi' by Perugino, depth profiles at two spots showed a thicker textile layer at the joint between two wooden boards than elsewhere
- The paintings 'The Dinner' and 'The Dance' from the 'Pipenpoyse Wedding' were made in 1616. One was restored and solvent cleaned, the other never restored. $T_{2\text{eff}}$ depth profiles through the paint and canvas regions reveal low $T_{2\text{eff}}$ throughout for the cleaned painting and higher $T_{2\text{eff}}$ for the original painting at deeper depth, suggesting differences in leaching of low molecular weight constituents by *solvent cleaning* and *evaporation*, respectively
- *Frescoes* are painted on wet *mortar* applied in strata of finer and finer grain size towards the surface. The *moisture distribution* across the mortar layers reports the manufacturing technology and skills of the trade from former times such as in the walls of the Villa of the Papyri in Herculaneum

6. Hyperpolarization

Polarization and spin order
Electron spin resonance
Polarization transfer from electrons: DNP
    Overhauser DNP
    DNP assisted by temperature
    DNP assisted by light: Diamonds and gases
Transfer of spin order: Para-hydrogen

© Springer Nature Switzerland AG 2019   143
B. Blümich, *Essential NMR*,
https://doi.org/10.1007/978-3-030-10704-8_6

The NMR Experiment

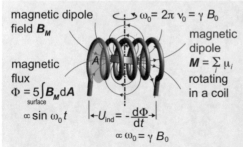

magnetic dipole field $\boldsymbol{B_M}$

$\omega_0 = 2\pi\,\nu_0 = \gamma\,B_0$

magnetic flux
$$\Phi = 5\int_{\text{surface}} \boldsymbol{B_M}\,\mathrm{d}\boldsymbol{A}$$
$$\propto \sin\omega_0 t$$

magnetic dipole
$$\boldsymbol{M} = \sum_i \boldsymbol{\mu}_i$$
rotating in a coil

$$\left|{-U_{\text{ind}} = -\frac{\mathrm{d}\Phi}{\mathrm{d}t}}\right|$$
$$\propto \omega_0 = \gamma\,B_0$$

Faraday's law of induction: The voltage U_{ind} induced in N closed loops equals the rate of change of the magnetic flux Φ enclosed by the loops

The flux is the magnetic field $\boldsymbol{B_M}$ times the areas $\boldsymbol{A}$ of the loops through which it passes

The magnetic field is generated by the sum $\boldsymbol{M}$ of all magnetic dipoles $\boldsymbol{\mu}$

$\boldsymbol{M}$ is proportional to the magnetic polarization

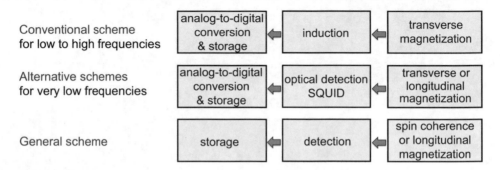

| Conventional scheme for low to high frequencies | analog-to-digital conversion & storage | ⬅ | induction | ⬅ | transverse magnetization |

| Alternative schemes for very low frequencies | analog-to-digital conversion & storage | ⬅ | optical detection SQUID | ⬅ | transverse or longitudinal magnetization |

| General scheme | storage | ⬅ | detection | ⬅ | spin coherence or longitudinal magnetization |

Nuclear Polarization and Sensitivity

- The *nuclear polarization* is the dipolar *longitudinal magnetization*
- It is given by the sum of population differences of energy levels between which transitions can be stimulated by radio-frequency photons
- Since a photon has spin 1, these energy levels differ in magnetic quantum number m by $|\Delta m| = 1$
- The polarization defines the signal amplitude M_0 after an excitation pulse provided all off-diagonal elements in the density matrix are zero before
- In *thermodynamic equilibrium* it is given by the *Boltzmann distribution*. For N spins, $M_0 = N\gamma\hbar\,\dfrac{\sum\limits_{m=-I}^{m=I} m\,\exp\{m\hbar\,\omega_0/k_B T\}}{\sum\limits_{m=-I}^{m=I} \exp\{m\hbar\,\omega_0/k_B T\}}$, and at high temperature, $M_0 \propto B_0$
- The *sensitivity of NMR* is specified by the *signal-to-noise ratio S/N*
- With *inductive detection* it is given by the electromotive force (voltage) induced in the detection coil relative to the standard deviation of the noise
- The induced voltage is proportional to the polarization and the time-derivative of the magnetic flux resulting from the sum of nuclear magnetic moments, which oscillates with the *Larmor frequency* $\omega_0 = \gamma\,B_0$ in the detection field B_0
- Therefore, the sensitivity is proportional to the product of the nuclear polarization and the strength B_0 of the detection field: $S/N \propto M_0\,B_0$

Improving the Sensitivity of NMR

Separate the polarization and detection processes

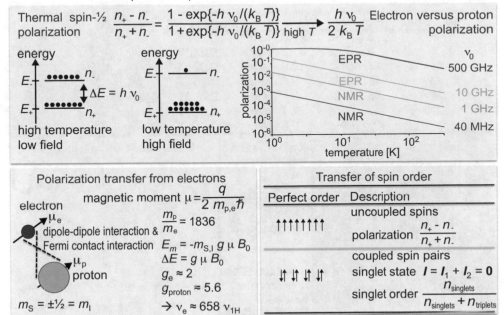

Thermal spin-½ polarization
$$\frac{n_+ - n_-}{n_+ + n_-} = \frac{1 - \exp\{-h\,\nu_0/(k_B T)\}}{1 + \exp\{-h\,\nu_0/(k_B T)\}} \xrightarrow{\text{high } T} \frac{h\,\nu_0}{2\,k_B T}$$
Electron versus proton polarization

energy

E_- •••••• n_-

$\updownarrow \Delta E = h\,\nu_0$

E_+ •••••• n_+

high temperature
low field

energy

E_- • n_-

E_+ •••••• n_+

low temperature
high field

Polarization transfer from electrons

electron
μ_e
dipole-dipole interaction &
Fermi contact interaction

μ_p
proton

$m_S = \pm\tfrac{1}{2} = m_I$

magnetic moment $\mu = \dfrac{q}{2\,m_{p,e}}\hbar$

$\dfrac{m_p}{m_e} = 1836$

$E_m = -m_{S,I}\, g\, \mu\, B_0$

$\Delta E = g\, \mu\, B_0$

$g_e \approx 2$

$g_{proton} \approx 5.6$

$\rightarrow \nu_e \approx 658\,\nu_{1H}$

Transfer of spin order

Perfect order	Description
	uncoupled spins
↑↑↑↑↑↑↑↑	polarization $\dfrac{n_+ - n_-}{n_+ + n_-}$
	coupled spin pairs
↕↑ ↕↑ ↕↑ ↕↑	singlet state $I = I_1 + I_2 = 0$
	singlet order $\dfrac{n_{singlets}}{n_{singlets} + n_{triplets}}$

Boosting the Sensitivity

- In ordinary NMR experiments single-quantum spin coherence, i.e. transverse magnetization is detected
- The conventional form of detection is *nuclear induction*
- If the spins are thermally polarized in the detection field, the *sensitivity* of induction NMR is roughly proportional to the square of the NMR frequency
- Alternative schemes separate the polarization and detection steps to optimize each individually
- At low frequencies the *longitudinal magnetization* (*polarization*) and the *transverse magnetization* can be detected with high sensitivity optically as well as with superconducting quantum interference devices (*SQUIDS*)
- The amplitude of the detected magnetization can be increased beyond the thermodynamic equilibrium value of the longitudinal magnetization by transferring spin order from sources with higher spin order
- For example, the Larmor frequency of electrons is 658 times higher than for ^{1}H
- At low temperature the *electron polarization* can approach 100 %
- It can be transferred to nuclei with *dynamic nuclear polarization* (*DNP*)
- For that unpaired electrons need to be available at sufficiently high density
- Another source of spin order is *para-hydrogen*. There, the two proton spins are aligned in antiparallel manner, producing a singlet state with zero total spin
- When breaking the symmetry of p-H_2, this spin order can be transferred to nuclear target spins by chemical reaction or through *spin-spin coupling*

EPR Equipment

EPR frequencies

Microwave band	Frequency	Wavelength	Magnetic field
S	3.0 GHz	100.0 mm	0.107 T
X	9.5 GHz	3.15 mm	0.339 T
K	23 GHz	13.0 mm	0.82 T
Q	35 GHz	8.6 mm	1.25 T
W	95 GHz	3.15 mm	3.3 T
D	140 GHz	2.1 mm	4.9 T
-	360 GHz	0.83 mm	12.8 T

Wikimedia:
public domain

Yevgeny Konstantinovich
Zavoisky - Евгений
Константинович Завойский,
1907 – 1976

Pulsed X to Q band EPR
spectrometer

X-band CW EPR
spectrometer

Some stable radicals for EPR

Courtesy of Rüdiger Eichel, FZ Jülich

Benchtop EPR
spectrometer
(Courtesy of
Magnettech GmbH,
Berlin)

Tempo

Tempol

Trityl

Totapol
(a biradical)

Electron Paramagnetic Resonance: EPR

- *EPR* for *electron paramagnetic resonance* and *ESR* for *electron spin resonance* are used synonymously
- EPR was discovered in 1944 by Yevgeny Konstantinovich Zavoisky in Kazan
- Independently the method was developed by Brebis Bleaney in Great Britain
- EPR is a magnetic resonance technique similar to NMR. It relies on the presence of *unpaired electrons*
- Stable unpaired electrons are rare in nature, so that stable free radicals are introduced into the sample as spin probes to report sample properties by EPR
- Whereas typical NMR frequencies employed for chemical analysis are in the radio-frequency regime between 40 and 1000 MHz, typical EPR frequencies are in the microwave regime between 3 - 500 GHz
- The higher resonance frequencies make EPR more sensitive than NMR
- EPR relaxation times are in the microsecond range and thus much shorter than NMR relaxation times, so that EPR lines are broad compared to NMR lines
- EPR spectra can cover bandwidths of several hundred Megahertz so that *pulsed excitation* is *frequency selective*
- Many EPR spectra are still measured today with the *continuous wave* technique by sweeping the magnetic field in such a way, that absorption spectra are obtained in derivative mode

One Electron Interacting with One Spin-½ Nucleus

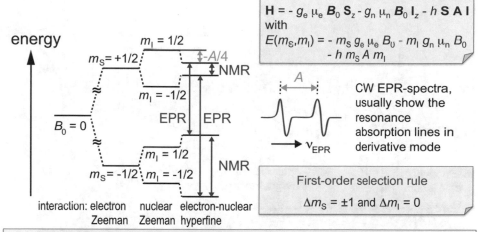

$$\mathbf{H} = - g_e\,\mu_e\,\mathbf{B}_0\,\mathbf{S}_z - g_n\,\mu_n\,\mathbf{B}_0\,\mathbf{I}_z - h\,\mathbf{S}\,\mathbf{A}\,\mathbf{I}$$
with
$$E(m_S, m_I) = - m_S\,g_e\,\mu_e\,B_0 - m_I\,g_n\,\mu_n\,B_0 - h\,m_S\,A\,m_I$$

CW EPR-spectra, usually show the resonance absorption lines in derivative mode

First-order selection rule

$$\Delta m_S = \pm 1 \text{ and } \Delta m_I = 0$$

In solid-state EPR, the first-order selection rule can be violated when the nuclear Zeeman and the hyperfine interactions are of similar magnitude. Then transitions with simultaneous I and S spin flips ($\Delta m_S = \pm 1$ and $\Delta m_I = \pm 1$) become weakly allowed, and the EPR signal is affected by the nuclear Zeeman and quadrupole interactions. As this effect occurs mainly in solids it is called the "solid effect"

Basic Theory

- The *Hamiltonian* of one electron spin interacting with nuclear spins k is given by
$$\mathbf{H}_{spin} = - g_e\,\mu_e\,\mathbf{B}_0\,\mathbf{S} - \Sigma\,g_{n,k}\,\mu_n\,\mathbf{B}_0\,\mathbf{I}_k - h\,\Sigma\,\mathbf{S}\,\mathbf{A}_k\,\mathbf{I}_k - h\,\Sigma\,\mathbf{I}_k\,\mathbf{Q}_k\,\mathbf{I}_k,$$

  | Zeeman int., electron spin | Zeeman int., nuclear spin | hyperfine interaction | quadrupole interaction |

  where μ_n is the nuclear magnetic moment, $\mathbf{I}_k$ the nuclear spin vector operator, $\mathbf{A}_k$ the electron nuclear hyperfine tensor in Hz, and $\mathbf{Q}_k$ the nuclear quadrupole coupling tensor in Hz
- Considering just one electron interacting with one spin ½ nucleus one obtains in the NMR notation
$$\mathbf{H}_{spin} = -\gamma_S\,\hbar\,\mathbf{S}\,\mathbf{B}_0 - \gamma_I\,\hbar\,\mathbf{I}\,\mathbf{B}_0$$
$$- \gamma_S\,\gamma_I\,\hbar^2\,(8\pi/3)\,|\Psi(0)|^2\,(\mathbf{I}\,\mathbf{S}) - \gamma_S\,\gamma_I\,\hbar^2\,[3\,(\mathbf{I}\,\mathbf{r})\,(\mathbf{S}\,\mathbf{r})\,/r^5 - \mathbf{I}\,\mathbf{S}\,/r^3].$$
  Terms 1 and 2 are the *Zeeman interactions* of the electron and the nucleus. Term 3 is the scalar *Fermi contact interaction* which corresponds to the *J-coupling* between nuclear spins in NMR. Here $\Psi(0)$ is the electron's wave function at the nucleus. Term 4 is the time- and distance-dependent direct *dipole-dipole interaction* between electron and nucleus
- The first-order selection rules for EPR transitions allow changes in *magnetic quantum numbers* $\Delta m_S = \pm 1$ and $\Delta m_I = 0$, so that the electron spin orientation changes but not the nuclear spin orientation
- The associated energy difference defines the hyperfine constant A in MHz

Overhauser DNP in a Two-Spin System

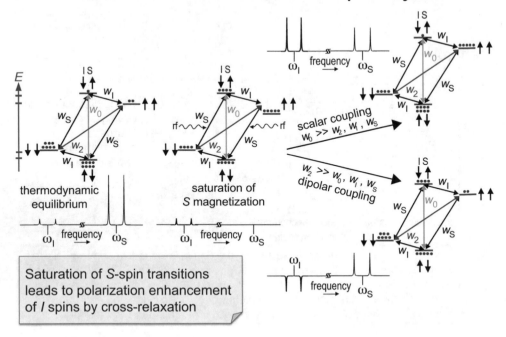

Saturation of S-spin transitions
leads to polarization enhancement
of I spins by cross-relaxation

Dynamic Nuclear Polarization: DNP

- *DNP* denotes the transfer of electron spin polarization to atomic nuclei by means of a coupling between electrons and nuclei
- In a system of two types of spins I and S interacting in a static magnetic field the populations of the energy levels depend on the spin polarizations, e.g. the polarization of the I spins depends on the polarization of the S spins
- When saturating the S-spin transitions by irradiation with frequency ν_S the differences in S-spin populations vanish
- If either the zero-quantum cross-relaxation rate w_0 or the double-quantum cross-relaxation rate w_2 is larger than the single-quantum relaxation rate w_1, the I-spin population differences increase
- This was predicted by A. Overhauser and is known as the *Overhauser effect*
- When the S spins are electrons and the I spins nuclei, the effect is termed *dynamic nuclear polarization,* in short: *DNP*
- In liquids, only *Overhauser DNP* by *cross-relaxation* is observed for electron und nuclear spins coupled indirectly by the the *Fermi contact interaction*
- In solids the direct dipole-dipole interaction enables further DNP mechanisms. These are the *solid effect*, the *cross effect*, and *thermal mixing*
- All of them may arise simultaneously and enhance the nuclear polarization
- DNP enables studies within hours that otherwise would take months to conduct and introduces contrast between species that otherwise appear alike

Making Use of Overhauser DNP

Theory: The *Overhauser-DNP enhancement* $E = M_{zl}/M_{0l} = 1 - \rho\, f\, s\, |\gamma_S|/\gamma_l$ is calculated from the relaxation rates w_i, where

   M_{zl}: current nuclear polarization (hyperpolarization)
   M_{0l}: nuclear thermodynamic equilibrium polarization
   $\rho = (w_2 - w_0)/(w_0 + 2w_1 + w_2)$ is the coupling factor
   $f = (w_0 + 2w_1 + w_2)/(w_0 + 2w_1 + w_2 + w^0)$ is the leakage factor
   $s = (M_{0S} - M_{zS})/M_{0S}$ is the saturation factor
   with
   M_{zS}: partially saturated electron polarization
   M_{0S}: thermodynamic equilibrium electronic polarization
   $w^0 = 1/T_1^{\,0}$ is the longitudinal nuclear relaxation rate without free radicals
The maximum enhancement is $E = 658$ for protons

The use of Overhauser-DNP

The *saturation factor s* approaches 1 for optimum EPR saturation

The *leakage factor* can be estimated from the NMR relaxation times T_1 and $T_1^{\,0}$ with and without free electron spins present, respectively, as $f = 1 - T_1/T_1^{\,0}$

Overhauser DNP experiments aim at determining the *coupling factor* ρ. It critically depends on the strength and fluctuation of the electron-nuclear interaction. From models based on spectral densities of motion, molecular translational diffusion coefficients can be determined in the vicinity of spin probes marking particular environments such as the surface water in proteins

Overhauser DNP

- The energy level diagram of an electron spin interacting with a nuclear spin ½ accounts for a difference in sign of the gyromagnetic ratios of particles with masses m and opposite charges q, $\hbar\,\gamma = g\,\mu = g\,q/(2m)$
- Saturating the electron transitions enhances the NMR signal if w_0 or $w_2 > w_1$
- In the limit of pure *dipole-dipole coupling* $w_2 = 6\,w_0$, and the NMR signal is enhanced negatively
- In the limit of pure *scalar coupling*, *cross-relaxation* proceeds only through w_0, and the NMR signal is enhanced positively
- The cross-relaxation rates w_0 and w_2 are large if energy between spins and lattice can be exchanged
- This is the case if the scalar or the dipolar electron-nuclear couplings are modulated through *translational* or *rotational motion* with an inverse *correlation time* τ (100 ps) similar to the electron resonance frequency (10 GHz at 0.35 T)
- Translational collisions of small molecules in liquids, e.g. have correlation times of about 100 ps, while the rotational motion of a complex of two small molecules can have correlation times on the order of 10 ps
- Therefore, the DNP enhancement of the liquid-state NMR signal at 10 GHz EPR frequency essentially reports *diffusion* of small molecules

The Solid Effect and the Cross Effect

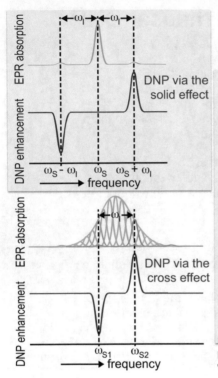

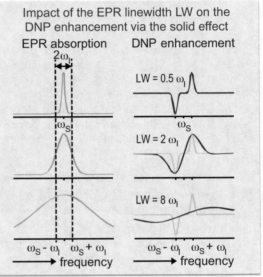

Impact of the EPR linewidth LW on the DNP enhancement via the solid effect

From B. Corzilius, **Phys. Chem. Chem. Phys. 18** (2016) 27190, Fig. 2 with permission CC BY 3.0

The Solid Effect

- Polarization is transferred, when coupled spins cross-relax or flip
- While the *Overhauser effect* relies on time-dependent interactions between electron and nuclear spins, other mechanisms rely on time-independent couplings and predominantly the direct coupling in solids
- The *solid effect* arises when the electrons are irradiated at frequency $\nu_{rf} = \nu_S \pm \nu_I$ with an offset of the nuclear resonance frequency ν_I above or below their resonance frequency ν_S to excite zero- or double-quantum transitions directly
- Depending on the offset, the nuclei are hyperpolarized with different signs
- These transitions are forbidden in first-order perturbation theory, but are weakly allowed in second order at high B_1 field and low B_0 field
- The transition probability and the *enhancement* scale with ν_I^{-2}. This dependence limits the use of the solid effect to low field
- Moreover, the electron-spin resonance-line width should be smaller than the NMR resonance frequency, so that only one of the two transitions $\nu_S \pm \nu_I$ is stimulated at a time to avoid cancellation of NMR signal amplitudes
- For an inhomogeneously broadened, wide EPR line, *hyperpolarization* is generated at both frequencies and signals interfere destructively
- Therefore, *stable radicals* like trityl are utilized which give rise to narrow lines
- Once nuclear polarization has been created locally it spreads across the sample volume by *spin diffusion*

Thermal Mixing

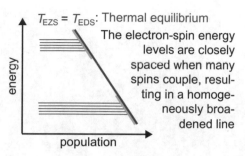

$T_{EZS} = T_{EDS}$: Thermal equilibrium

The electron-spin energy levels are closely spaced when many spins couple, resulting in a homogeneously broadened line

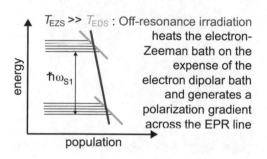

$T_{EZS} \gg T_{EDS}$: Off-resonance irradiation heats the electron-Zeeman bath on the expense of the electron dipolar bath and generates a polarization gradient across the EPR line

$\hbar\omega_{S1}$

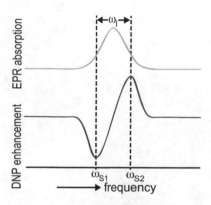

The nuclear Zeeman bath is cooled by the electron dipolar bath via energy conserving three-spin transitions

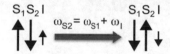

$S_1 S_2 I$ $\omega_{S2} = \omega_{S1} + \omega_I$ $S_1 S_2 I$

The Cross Effect and Thermal Mixing

- The cross effect and thermal mixing are three-spin processes of two electron spins and one nuclear spin satisfying the matching condition $\omega_{S2} - \omega_{S1} = \omega_I$
- Other than the *solid effect*, both effects arise from irradiating allowed transitions
- The *cross effect* requires an EPR resonance inhomogeneously broadened by *g* anisotropy with homogeneous linewidths smaller than ω_I
- The difference in electron Larmor frequencies provides the energy for the nuclear spin flip; biradicals improve the DNP efficiency
- With MAS the cross effect is currently the most efficient DNP mechanism
- Under MAS the energy levels shift periodically, and the polarization is transferred by reoccurring *level anti-crossings*
- *Thermal mixing* requires an EPR resonance with a linewidth on the order of the NMR frequency, homogeneously broadened from many interacting electrons
- The electron-electron coupling provides the energy for the nuclear spin flip
- Thermal mixing is explained in terms of spin temperature equilibrating between three interacting baths, the electron Zeeman system EZS, the electron dipolar system EDS, and the nuclear Zeeman system NZS
- Weak off-resonance irradiation produces a polarization gradient across the EPR line, equivalent to cooling the EDS
- Being in contact with the EDS bath, the NZS bath is then cooled by energy conserving three-spin processes leading to DNP enhancement
- Although the enhancements by the cross effect and thermal mixing scale with B_0^{-1}, both are successfully used at high field

Practice of Dissolution DNP

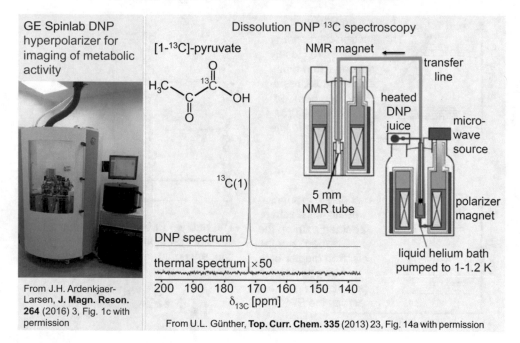

GE Spinlab DNP hyperpolarizer for imaging of metabolic activity

Dissolution DNP ^{13}C spectroscopy

[1-^{13}C]-pyruvate

NMR magnet

transfer line

heated DNP juice

micro-wave source

^{13}C(1)

5 mm NMR tube

polarizer magnet

DNP spectrum

thermal spectrum |×50

liquid helium bath pumped to 1-1.2 K

200 190 180 170 160 150 140
δ_{13C} [ppm]

From J.H. Ardenkjaer-Larsen, **J. Magn. Reson. 264** (2016) 3, Fig. 1c with permission

From U.L. Günther, **Top. Curr. Chem. 335** (2013) 23, Fig. 14a with permission

The Use of DNP

- The largest *DNP enhancement* is achieved at high field and low temperature
- Depending on type of nucleus and field strength, three to five orders of magnitude in polarization can be gained by freezing liquid samples with the free radicals into a glass
- *DNP juice*, a mixture of 60/30/10 (vol%) d_8-glycerol/D_2O/H_2O is a popular solvent for the stable radicals and the target molecules. It forms a glass upon cooling and avoids *crystallization* damage in particular to biological materials
- The polarization step takes minutes to tens of minutes
- Often protons are hyperpolarized first and then polarization is transferred to other nuclei
- Low sample concentrations can be analyzed in single scans by MAS that otherwise would require weeks and months of signal averaging
- With *dissolution DNP*, on the other hand, the hyperpolarized, glassy sample is rapidly melted for analysis by liquid-state NMR spectroscopy or MRI
- This temperature-jump method yields polarization levels in the two-digit percent range
- DNP hyperpolarized *pyruvate* injected into living organisms accumulates in regions of high metabolic activity such as cancerous tissue and lightens up in MRI
- Tumor cells convert pyruvate into lactate, while healthy cells produce bicarbonate
- Spectroscopic ^{13}C MRI detection of selectively ^{13}C enriched and hyperpolarized pyruvate reveals details of the pyruvate digestion into metabolites
- Commercial instruments such as General Electric's *Spinlab* or Oxford Instruments' *Hypersense* produce biocompatible solutions of DNP hyperpolarized tracer molecules for molecular imaging in minutes with a 10.000 fold gain in sensitivity

Defects in Diamond

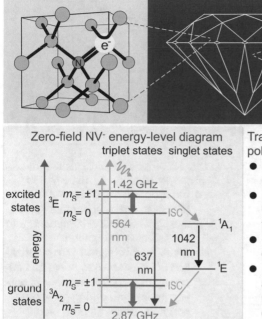

NV⁻ center

P1 center

Zero-field NV⁻ energy-level diagram
triplet states singlet states

1.42 GHz

excited states ^3E $m_S= \pm1$ $m_S= 0$

ISC

$^1A_1$

564 nm

637 nm

1042 nm

^1E

energy

ground states $^3A_2$ $m_S= \pm1$ $m_S= 0$

ISC

2.87 GHz

Transfer of optically induced electron polarization to nuclei

- 90% electron spin polarization can be gained in less than 1 μs
- It is transferred to ^{13}C nuclei by Overhauser DNP, the solid effect, and through level anti-crossings at low field
- ^{13}C polarization spreads across the diamond lattice by spin diffusion
- Hyperpolarized nano-diamond powders are envisioned as biocompatible ^{13}C markers in MRI and as polarization sources for nuclei in liquid molecules

DNP from Nitrogen Vacancy Centers in Diamond

- High thermal electron polarization results from high field and low temperature
- High electron polarization can also be created at low field and high temperature by *optical pumping* of crystal defects such as NV⁻ and P1 defects in *diamond*
- Uncharged *nitrogen vacancy* (NV) defects in diamond occupy two neighboring lattice sites, one being occupied by a nitrogen atom and the other being vacant
- The vacant site has three unpaired electrons in rapid exchange. Two form a quasi covalent bond via the lone pair of the nitrogen, and one electron can pair with a negative charge resulting in an *NV⁻ defect* with electron spin $S = 1$
- In a diamond *P1 defect* an N Atom substitutes a C atom, leaving one $S = \frac{1}{2}$ unsaturated electron spin
- The optical energies of the NV⁻ center are split by the electron spin states
- When excited with a green *laser* at higher than the transition energy, phonons are generated, which enhance the non-radiative depopulation of the $m_S = \pm1$ excited states, whereas the $m_S = 0$ state depopulates mainly radiatively
- The populations of the $m_S = \pm1$ excited states preferably relax by inter-system crossing (ISC) to the $m_S = 0$ ground state so that it's population is increased preferentially, and the electron-spin states can be detected optically
- In a magnetic field, the *Zeeman interaction* shifts the energy levels. They are further split by the electron-nuclear hyperfine interaction
- Whereas the $\Delta m_S = 1$ transitions depend on the diamond orientation, the $\Delta m_S = 2$ transition does not, so that it lends itself for DNP in *diamond powder*

Hyperpolarization of Xenon Gas

Resonant irradiation with circularly polarized light

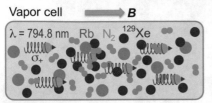

Vapor cell   ⟹ *B*

λ = 794.8 nm   Rb N_2 ^{129}Xe σ_+

Optical hyperpolarization of Rb electron spins

collisional mixing

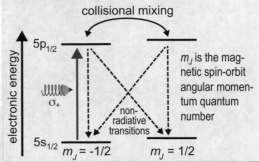

m_J is the magnetic spin-orbit angular momentum quantum number

5p$_{1/2}$

non-radiative transitions

5s$_{1/2}$ m_J = -1/2 m_J = 1/2

electronic energy

Hyperpolarization of ^{129}Xe nuclear spins by Fermi contact with optically polarized Rb electron spins in collision complexes

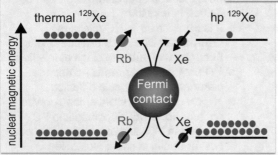

thermal ^{129}Xe   hp ^{129}Xe

Rb   Xe

Fermi contact

Rb   Xe

nuclear magnetic energy

Hyperpolarization of noble gases
- Electron spins in alkali metal vapor are polarized by spin-state selective light excitation competing with nonselective relaxation
- Alkali-metal electron-spin polarization is transferred to noble gas nuclei by the electron-nuclear hyperfine interaction during gas-phase collisions between both atoms

Hyperpolarization of Noble Gases

- The nuclear spins of *noble gas atoms* like ^{129}Xe, ^{83}Kr and ^{3}He can be hyperpolarized by *Spin-Exchange Optical Pumping* (*SEOP*)
- They serve to image void spaces like the lung, and trapped in cages chemically linked to marker molecules report the host chemistry by their chemical shift
- Most commonly ^{129}Xe (I = 1/2) is hyperpolarized. It's natural abundance is 26%
- In 1949 Alfred Kastler (1966 Nobel Prize in Physics) showed that the electron spins in alkali metal vapors can be polarized with circularly polarized *laser* light
- Subsequently Happer showed that the electron-spin order can be transferred to the nuclear spins of ^{3}He and ^{126}Xe gas mixed with the alkali vapor
- Today large quantities of *noble gases* can be hyperpolarized up to nearly 100%
- By irradiating *rubidium vapor* resonantly with σ^+ polarized light, conservation of angular momentum selectively depletes the ground-state m_J = -1/2 population
- Rb collisions with gas atoms or molecules equalize both excited-state m_J = ±1/2 populations and repopulate both m_J = ±1/2 ground-states at similar rates
- Collisions with N_2 buffer gas quench stimulated Rb emissions
- Continual light exposure then builds up Rb electron-spin polarization
- The polarization is kept by a weak magnetic field in direction of the laser light
- It is transferred from Rb electrons to ^{129}Xe nuclei via the *hyperfine interaction* in collisions assisted by N_2 forming long-lived van-der-Waals complexes
- The exceptionally wide Xe *chemical shift* range of 8000 ppm helps to identify different chemical and physical environments

Generation of p-H$_2$

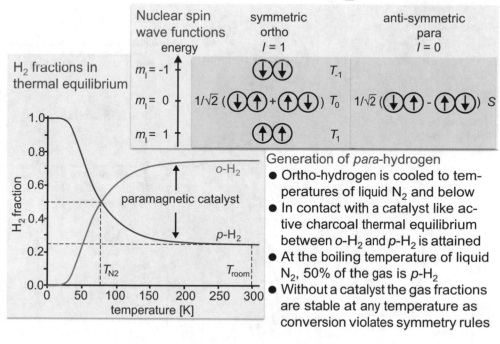

Generation of *para*-hydrogen
- Ortho-hydrogen is cooled to temperatures of liquid N_2 and below
- In contact with a catalyst like active charcoal thermal equilibrium between o-H$_2$ and p-H$_2$ is attained
- At the boiling temperature of liquid N_2, 50% of the gas is p-H$_2$
- Without a catalyst the gas fractions are stable at any temperature as conversion violates symmetry rules

Para-Hydrogen

- *Para-hydrogen* provides nuclear *spin order* that can be transferred to molecules by chemical reaction and by physical contact
- The chemical pathway is known as *PHIP* (*Para-Hydrogen Induced Polarization*) and the physical pathway as *SABRE* (*Signal Amplification By Reversible Exchange*)
- The two proton spins of the hydrogen atom can be paired in four ways and grouped according to symmetry under exchange of spins
- In *ortho*-hydrogen both spins are parallel with total nuclear spin $I = 1$, and the nuclear spin wave function is symmetric. In p-H$_2$ the nuclear magnetic moment is zero, so $I = 0$, and the wave function is anti-symmetric
- At room temperature the four states of hydrogen are about equally probable
- Due to symmetry of the H$_2$ wave function including nuclear spin states and molecular rotation, p-H$_2$ is more abundant in thermal equilibrium at low temperature
- Because transitions between different spin symmetries are forbidden, o- and p-H$_2$ fractions equilibrate only with assistance of a symmetry breaking *catalyst*
- *Para*-H$_2$ is generated from o-H$_2$ by passing it over a catalyst at low temperature
- Free from *catalyst* it can be stored at room temperature for long time
- Since the nuclear spin of p-H$_2$ is zero, it does not produce any NMR signal
- PHIP and SABRE break the symmetry of p-H$_2$ and generate NMR signal

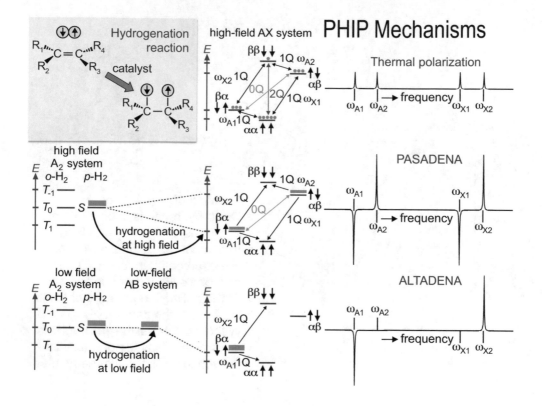

Para-Hydrogen Induced Polarization: PHIP

- *PHIP* denotes a hydrogenation reaction with p-H_2 where the *spin order* is preserved in the product but the magnetic equivalence is lifted
- Depending on whether the reaction proceeds at low or high magnetic field, different energy levels are populated
- The reaction at high field in the weak coupling limit is known as *PASADENA* (Para-Hydrogen and Synthesis Allow Dramatic Nuclear Alignment)
- Here the A_2 spin system of the two equivalent proton spins from p-H_2 is transformed into an AX system, whereby only $m = 0$ states are populated but the magnetic degeneracy is lifted, resulting in two anti-phase doublets
- The reaction at low field in the strong coupling limit is known as *ALTADENA* (Adiabatic Longitudinal Transport After Dissociation Engenders Net Alignment)
- At low field the *J*-coupling is large compared to the chemical shift difference, so that the spin system is classified as an AB system
- A 45° excitation pulse generates a single line with maximum intensity
- Following transfer of the sample to high field, the AB system becomes an AX system, and only two of the four single-quantum transitions are observed
- PHIP is used to study hydrogenation reactions and to produce hyperpolarized gases like propane from propyne for MRI of voids
- PHIP and SABRE rely on chemical expertise for *hyperpolarization* in identifying *catalysts* suitable for the reaction and the complex formation, respectively, while DNP requires extensive equipment resources

Spin-Order Transfer from p-H$_2$

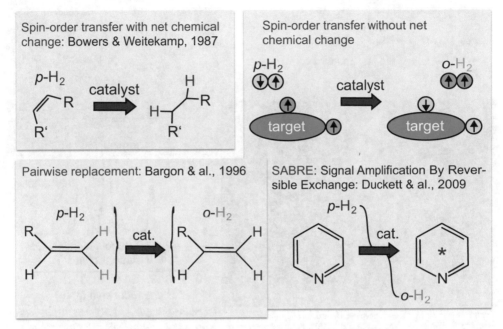

Spin-order transfer with net chemical change: Bowers & Weitekamp, 1987	Spin-order transfer without net chemical change
Pairwise replacement: Bargon & al., 1996	SABRE: Signal Amplification By Reversible Exchange: Duckett & al., 2009

More Ways of Tapping the p-H$_2$ Spin Order

- The hydrogenation PHIP involves a reaction with p-H$_2$ as one educt, producing a product in which the two hydrogen atoms appear at different chemical shifts so that their former magnetic equivalence in the H$_2$ molecule is lifted
- Other ways of transferring spin order from p-H$_2$ do not change the net chemistry of the molecule targeted for hyperpolarization
- One way is the pairwise *chemical replacement* of two hydrogens in the target molecule by the two singlet p-H$_2$ atoms
- Another way is the *physical transfer of spin order* from a p-H$_2$ molecule to a target molecule during the lifetime of a temporary complex between both molecules by means of *spin-spin coupling* between both molecules
- This non-hydrogenative polarization transfer is termed *SABRE*
- To extend the lifetime of collision complexes in solution, *polarization transfer catalysts* are employed, which are inefficient in the chemical sense of lowering the energy of the transition state for a chemical transformation
- The key to polarizing a large class of molecules by SABRE is competence in chemical catalysis
- The challenge to master is to maintain the *spin order* during transfer within the lifetime of the complex, in particular in protic solvents like life-sustaining water, where protons exchange rapidly

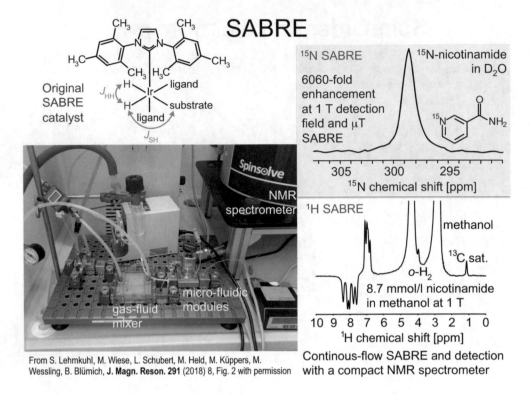

From S. Lehmkuhl, M. Wiese, L. Schubert, M. Held, M. Küppers, M.
Wessling, B. Blümich, **J. Magn. Reson. 291** (2018) 8, Fig. 2 with permission

Continous-flow SABRE and detection
with a compact NMR spectrometer

Polarization Transfer by SABRE

- Typical *SABRE polarization transfer catalysts* are Ir^{III} complexes bearing bulky electron-donating ligands that contain *cis*-hydrides and three coordinated ligands
- The ligands are solvent molecules or substrate molecules like nitriles, diazirines, pyridine, and other molecules containing aromatic nitrogen atoms
- The equatorial *cis*-hydrides and ligands exchange rapidly between p-H_2 and o-H_2 as well as between polarized and unpolarized substrate molecules
- *Spin-order* from p-H_2 to unpolarized substrate is transferred coherently through the *J-coupling* network acting during the lifetime of the complex
- *Transfer paths* are enabled by matching transition energies of hydride and substrate nuclei
- For ^1H target nuclei optimum matching fields are around 6.5 mT; for heteronuclei like ^{15}N they are in the µT regime below the earth magnetic field of 50 µT
- A wide range of analytes can be hyperpolarized by relaying the p-H_2 spin order through exchangeable protons of hyperpolarized intermediates
- Even water can be hyperpolarized although it's hydrogen solubility is low
- As with DNP, substrates for use as biomarkers need to be free of *catalyst*
- Flow *hyperpolarization* with continual replacement of o-H_2 by p-H_2 enables signal averaging and considerable degrees of experimental freedom for detecting trace concentrations from analytes

Correction to: Essential NMR

Correction to:
B. Blümich, *Essential NMR*,
https://doi.org/10.1007/978-3-030-10704-8

In the original version of the book, the following belated corrections have been made:

In chapter 1 (p. 6), the author name "A. Haber-Pohlmeier" has been changed to "S. Haber-Pohlmeier".

In chapter 3 (p. 64), the horizontal axis of the graph has been corrected.

In chapter 4 (p. 92), "From A. Haase, J. Frahm, D. Matthei, W. Hanicke, K. D. Merboldt, J. Magn. Reson. 67 (1986) 258 with permission" has been changed to "From A. Haase, J. Frahm, D. Matthei, W. Hanicke, K. D. Merboldt, J. Magn. Reson. 67 (1986) 258, Fig. 1 with permission"; (p. 97), "From P. Prado, B. J. Balcom, M. Jama, J. Magn. Reson. 137 (1999) 59 with permission" has been changed to "From P. Prado, B. J. Balcom, M. Jama, J. Magn. Reson. 137 (1999) 59, Fig. 1 with permission".

In chapter 5 (p. 114), "The transverse magnetization $M_{xy}(t) = M_0 \int P(R) \exp\{-i\,q\,R\}\,dR$ of an ensemble of spins diffusing different distances R during the diffusion time is determined by the Fourier transform of the probability density P(R) of displacements, which is known as the propagator in NMR" has been changed to "The transverse magnetization $M_{xy}(t) = M_0 \int P(R) \exp\{-i\,q\,R\}\,dR$ of an ensemble of spins diffusing

The updated version of this book can be found at
https://doi.org/10.1007/978-3-030-10704-8

© Springer Nature Switzerland AG 2019
B. Blümich, *Essential NMR*,
https://doi.org/10.1007/978-3-030-10704-8_7

different distances R during the diffusion time without relaxation is determined by the Fourier transform of the probability density P(R) of displacements, which is known as the propagator in NMR"; (p. 117), the incomplete figure has been corrected; (p. 132), the inline equation at the end of the sentence starting "The normalized integral of the..." has been corrected; (p. 133), one of the figures has been corrected; and (p. 134), the author name "Ohligschläger" has been changed to "Oligschläger".

Index

absorption signal 24
acceleration 77, 80, 106, 108
acquisition time 85
aging 136, 141
aging, chemical 141
aging, physical 141
aging, thermo-oxidative 136, 141
aging, UV 136
alignment echo 49
ALTADENA 156
amorphous domains 138, 140
angle dependence of the NMR frequency 50
angular momentum 13
anisotropy of the chemical shift 51, 64
anisotropy parameter 38
asymmetry 40
asymmetry parameter 38
attenuation function 76, 83
auto-correlation function 82
auto-correlation function, pore shape 117
bicycle wheel 13
Bloch equation 16
Boltzmann distribution 46, 144
bound fluid 33, 113
buoyancy 123
calibration curves 137
catalyst 155, 156, 158
Chang model 106
chemical analysis 2
chemical engineering 2, 100
chemical replacement 157
chemical shielding 39, 40, 42
chemical shift 5, 12, 55, 74, 108
chemical shift dispersion 69

chemical shift resolution 61
chemical structure 11
Coates-Timur model 125
coherence order 98
coherence, double-quantum 58, 61, 69, 98
coherence, multi-quantum coherence 43, 53, 57, 68, 71, 98
coherence, single-quantum 43, 53, 98
coil 18
commutator 46
connectivity 57
conservation agent 142
continuous wave 30, 146
contrast 93, 95, 134
coordinate frame, laboratory 15, 19
coordinate frame, rotating 19, 21, 25, 46, 76
coordinates, Cartesian 84
coordinates, cylindrical 84
correlation NMR 54
correlation NMR, D-T_2 130
correlation NMR, T_1-T_2 130
correlation spectroscopy 57
correlation time 48, 149
corrosion 141
COSY 57, 58, 59, 60, 61, 69
coupling, direct 44
coupling, dipole-dipole 149
coupling, hetero-nuclear dipole-dipole 50, 51
coupling, indirect 44, 59
coupling, indirect spin-spin 5, 42, 44
coupling, J 44, 59, 147, 158
coupling, residual dipole-dipole 65, 135
coupling, scalar 149
coupling, spin-spin 145, 157
coupling factor 149

coupling tensor 36
CP 51
CPMG decay 126, 137
CPMG echo envelope 115, 132
CPMG echo train 116, 122
CPMG sequence 27, 114
CRAMPS 52
cross effect 148, 151
cross-filtration 108
cross polarization 5, 51
cross-relaxation 65, 66, 148, 149
cross-relaxation rate 65
crosslink density 135, 137
crude oil 125, 128
crystalline domains 138
crystallinity 138, 139, 140
crystallization 141, 152
crystallization, solvent-induced 141
crystallization, strain-induced 135
crystallization, temperature-induced 141
cultural heritage 142
curing time 136
curvature 77
dead oil 127, 128
deadtime 24, 49, 121, 123
degradation 141
density matrix 45, 52, 57
density matrix equation of motion 46
DEPT 68
detection period 58, 89, 116
detection time 66
deuteron wideline NMR spectroscopy 48
diamond 153
diamond powder 153
diffraction 82, 117
diffusion 77, 114, 115, 116, 133
diffusion, restricted 105, 117, 133

© Springer Nature Switzerland AG 2019
B. Blümich, *Essential NMR*,
https://doi.org/10.1007/978-3-030-10704-8

diffusion, spin 64, 150
diffusion, translational 105
diffusion coefficient 112
diffusion coefficient,
     apparent 113
diffusion equation 123
diffusion filter 116
diffusion in internal
     gradients 123
diffusion length, mean 117
diffusion of methane 128
diffusion of water 128
diffusion regime, slow 123
diffusion time 114
diffusion weight 126
diffusion-ordered
     spectroscopy 129, 131
digital resolution 95
dipolar decoupling 50, 51
dipolar filter 53
dipole moment 36
directional drilling 120
dispersion signal 24
displacement 103, 106, 114
displacement, coherent 105
displacement, incoherent
     105
displacement, probability
     density 114
displacement distribution
     117
displacements, 2D
     distributions 105
distortionless enhancement
     by polarization transfer
     68
distribution 62, 103
distribution, isotropic 47
distribution, temperature 94
distribution of accelerations
     106, 108
distribution of amplitudes 32
distribution of displacements
     103, 105
distribution of orientation
     angles 62
distribution of orientations
     48
distribution of relaxation
     rates 32, 33

distribution of relaxation
     times 122
distribution or reorientation
     angles 62
distribution of velocities 103
distributions of diffusion
     coefficients 130
distributions of apparent
     diffusion coefficients 113
distributions of frequencies
     17, 32
distributions of relaxation
     rates 17, 32, 33, 113
distributions of relaxation
     times 32, 33, 126, 130,
     134, 137
DNP 145, 148
DNP, dissolution 152
DNP enhancement 152
DNP enhancement,
     Overhauser 149
DNP juice 152
DNP, Overhauser 148
DOSY 129, 131
double-quantum filtered
     imaging 98
double-quantum imaging
     98
drilling mud 127
droplet-size distribution 133
dynamic multi-dimensional
     NMR spectroscopy 62
dynamic nuclear
     polarization 145, 148
earth-field tool 121
echo 26, 49, 67
echo, spin 55, 90
echo, stimulated 26, 49,
     81, 116
echo amplitude 26
echo attenuation 116
echo planar imaging 86,
     96
echo time 26, 90, 126
editing of spectra 44
effective field 25
effective gradient 81, 115,
     117
effective relaxation time 115
eigenfunction 45

eigenvalue 39, 41, 45
eigenvector 41, 45
elastic modulus 137
elastomers 93, 135
elastomers, technical 135
electric oscillator 18
electrical resistivity 126
electromagnet 119
electron paramagnetic
     resonance 146
electron polarization 145
electron spin resonance
ellipse 38
emulsions 133
energy levels 52
enhancement 150
enhancement factor 65
ensemble average 46
EPI 86, 96
EPR 119
Ernst angle 92
ESR 119
Euler angles 41
evaporation 141
evolution period 58, 89,
     129
evolution time 54, 59,
     66, 85
exchange cross peaks 63
exchange NMR, 2D 5
exchange NMR, T_2-T_2 130
exchange NMR
     spectroscopy 62
exchange spectrum, 2D
     63
excitation power 30
expectation values 46
EXSY 58, 62, 66
fast diffusion limit 113, 123
fast imaging 86
fast motion limit 64
FID 17, 23, 59
field gradient 81, 113
field-gradient vector 75,
     76
field inhomogeneity 123
finite difference
     approximation 80, 104
FLASH imaging 92, 96
flow 106

flow, coherent 77, 115
flow, laminar 108, 116
flow encoding 101, 102
flow imaging 83
flow through a circular pipe,
   laminar 103
fluid, producible 33, 113,
   122, 125, 126
fluid content, volumetric 132
fluid saturation 113
fluid typing 113
foam density 135
FONAR 99
Fourier imaging 84
Fourier imaging, 3D 91
Fourier NMR 5, 23, 32, 84
Fourier NMR, 2D 129
Fourier NMR, multi-
   dimensional 109
Fourier pair 78
Fourier transform 24, 25,
   87
Fourier transformation 17,
   22, 23
Fourier transformation, 2D
   54, 55, 59
Fourier transformation, 3D
   91
Fourier transformation,
   multi-dimensional 84,
   109
Fourier-conjugate variables
   78, 109
Fourier transform of the
   measured signal 109
Fredholm equation 31
free fluid 121
free induction decay 17
free precession 46
frequency bandwidth 28
frequency encoding 85,
   89
frequency-exchange
   spectroscopy 106
frequency selective 119
frequency-selective pulse
   87
frescoes 142
gamma-ray scattering 122,
   126

Gauss function 117
Gaussian distribution 105,
   235
geophysical well logging 2
glass temperature 135, 137
gradient 31, 77
gradient, internal 113, 130
gradient coil 74, 75
gradient echo 81, 89,
   106, 114
gradient-echo imaging 92,
   93
gradient field 4, 21, 74,
   75, 76, 81
gradient-field vector 75
gradient modulation
   function, moments 78,
   83
gradient moment 79
gradient switching time 89
gradient-pulse pair, anti-
   phase 114
Hahn echo 26, 27, 49, 53,
   89, 112
Halliburton 120
Halliburton/Numar tool 121
Hamilton operator 45, 52
Hamiltonian 147
Hartmann-Hahn condition
   51
HETCOR 69, 70
heterogeneity, material 136
hetero-nuclear 2D NMR
   71
HMBC 69, 71
homo-nuclear decoupled
   1D spectrum 56
HSQC 69, 71
human tissue 134
hydrocarbon type 122
hydrogen index 123, 127
hydrogen index, apparent
   127
hyperpolarization 5, 150,
   156, 158
Hypersense 152
image 32
image, parameter 94, 99
image, parameter-weighted
   spin-density 94

image, spin-echo 92
image, T_2 94
image contrast 90
images, 3D 91
images, spectroscopic 95
imaging 4, 83
imaging, back-projection
   84, 86, 93
imaging, clinical 92
imaging, line-scan 86
imaging, spectroscopic 85
imaging, spin-echo 86,
   89, 90
imaging, spin-warp 85, 99
imaging, spiral 96
imaging of solids 85
impulse response 59
INADEQUATE 58, 61
INADEQUATE, 2D 69
indirect coupling, hetero-
   nuclear 67, 69
indirect method 133
inductive detection 114
inhomogeneous field 112,
   114
inhomogeneous field,
   magnetic 27
inside-out NMR 120
interaction, anisotropic 39,
   50
interaction, dipole-dipole 36,
   37, 40, 44, 52, 53, 64,
   66, 147, 149
interaction, electric
   quadrupole 40
interaction, Fermi contact
   147, 148
interaction, hyperfine 154
interaction, quadrupole 36,
   39
interaction, nuclear
   quadrupole 48
interaction, spin 42
interaction, spin rotation 64,
   128
interaction, symmetric 40
interaction, Zeeman 39, 42,
   147, 153
interaction anisotropy 39, 40,
   52

interaction ellipsoid 41
interaction energy 36, 43, 45
interaction tensor 47, 48
interference fringes 82
inverse detection 66, 69
inverse Laplace transformation 17, 32, 112, 122
inverse Laplace transformation, 1D 129
inverse Laplace transformation, 2D 116
inverse Laplace transformation, 2D algorithm 129
inversion 28, 65
inversion recovery 28
inversion recovery sequence 112
Jackson tool 121
k space 86
Kenyon model 125
Laplace methods, multidimensional 122
Laplace NMR 32, 112
Laplace NMR, 2D 126, 129
Laplace NMR, two-dimensional 130
Larmor frequency 13, 82, 144
laser 153, 154
leakage factor 149
level anti-crossings 151
linear field profile 78
linear magnetic field 80
linewidth 22
live oil 127
logging while drilling 120, 126
long-lived spin states 5
longitudinal magnetization 15, 22, 26, 28, 53, 64, 76, 88, 123, 144, 145
longitudinal relaxation 90
longitudinal relaxation time 16, 22
magic angle 40, 47, 50, 52
magic angle spinning 5, 50
magic echo 49
magnet, center-field 119
magnet, Halbach 119
magnet, stray-field 119

magnetic field 3
magnetic field distortions 93
magnetic field inhomogeneity 93
magnetic quantum number 14, 43, 147
magnetic resonance imaging 3, 74
magnetic shielding 12, 39
magnetization filters 97
magnetization, nuclear 15
MAS 50, 51, 52
material properties, average 136
materials science 2, 100
materials testing 4
mechanical properties 140
medical diagnostics 2
mercury intrusion porosimetry 124
methane 127
mixing period 58, 66, 129
mixing propagator 53
mixing time 62, 66, 130
moisture distribution 142
molecular dynamics 48
molecular motion 48
molecular motion, slow 48
molecular orientation 47, 48
molecular weight distribution 128
mortar 142
motion, rotational 64, 149
motion, translational 64, 149
MRI 3, 74, 134
multi-pulse NMR 52
multi-pulse sequence 52
multi-quantum build-up curves 53
multi-quantum coherence order 53
multi-quantum evolution period 53
multi-quantum experiment 58
multi-quantum pulse sequence 57
multiplet 33, 55

multiplet splitting 44, 47, 59
multiplex advantage 30
neutron scattering 122, 126
nitrogen vacancy 153
NMR 2
NMR, multi-dimensional 30, 33
NMR echo 5
NMR frequency 47, 52
NMR imaging 5, 136
NMR logging 4
NMR logging tool 122
NMR-MOUSE 4, 119, 132, 134, 136
NMR relaxometer 4
NMR relaxometry 3
NMR spectrometer 4
NMR spectroscopy 3, 74
NMR spectroscopy, high-resolution 129
NMR spectroscopy, multi-dimensional 43, 54, 129
NMR spectrum 12, 95
NMR spectrum, high-resolution 33
NMR spectrum, multi-dimensional 57, 66
NMR tomograph 4
NMR tomography 3
noble gases 154
NOE 62, 65
NOESY 58, 66
noise excitation 30
non-destructive testing 4
nuclear induction 145
nuclear Overhauser effect 62, 66
NV⁻ defect 153
off-set field 25, 76
oil, synthetic 127
optical pumping 153
orientation dependence 37
orientation dependence of the NMR frequency 52
orientation-dependent line splittings 43
oriented solids, partially 48
oscillation, forced 29, 30

oscillation, free 29
oscillator 21
Overhauser effect 65, 148, 150
OW4 112
P1 defect 153
paintings, stratigraphy 142
Pake spectrum 47, 48
para-hydrogen 145, 155
para-hydrogen induced polarization 155
PASADENA 156
Patterson function 82
Pauli spin matrices 46
peak integrals 33
permeability 113, 125, 126
permeability estimates 122
PFG NMR 79, 81
phase 23
phase, magnetization 77
phase, precession 76, 115
phase correction 24
phase cycling 71
phase encoding 85, 89
phase-encoding dimension 93
phase-encoding gradient 91
phase-encoding imaging 93
phase of the transverse magnetization 77, 78
phase shift of the echo 114
phase twist 55
PHIP 155, 156
Planck's constant 13
plants 134
point-spread function 83
polarization 145
polarization, magnetic 15
polarization, nuclear 144
polarization, time 126
polarization, transfer 67
polarization transfer, coherent 67, 70, 71
polarization transfer, hetero-nuclear 44
polarization transfer, incoherent 67
polarization transfer catalysts 157

polymer products 139
population differences 67
pore connectivity 122
pore size 113, 124
pore-shape function 118
pore-size distribution 113
pore-throat distribution 124
porosity 113, 121, 122, 123, 127
porous media 113, 130
position 77
position-exchange experiment 107
position-exchange NMR 108
position-exchange spectroscopy 106
powder spectrum 47, 50
POXSY 106
precession frequency 13, 76
preparation period 53, 58, 84, 129
preparation propagator 53
pressure, capillary 124
principal axes frame 41
principal value 38
probability densities 103
producible fluid 33
projection 31, 87, 104, 109
projection – cross-section theorem 102, 109
propagator 103, 105, 114, 117, 118
pulse 46
pulse, rf 25
pulsed excitation 30, 146
pulsed field-gradient NMR 79, 81
pulsed gradient fields 81
pyruvate 152
quadratic magnetic field 80
quadrature component 21
quadrupolar spins 98
quality control of rubber products 136, 137
quantum mechanics 13, 43, 45, 52
quantum number, spin 14
racetrack echo 26
radio frequency 18

RARE 96
receiver 21
recovery period 58, 90, 129
recycle delay 92
recycle period 58
relaxation 26, 115
relaxation, bulk 127
relaxation super-operator 46
relaxation time 26, 64, 112, 124, 135
relaxation time, average 132
relaxation time, bulk 123
relaxation time, logarithmic mean 128
relaxation-time cut-off 125
relaxation-time distribution 124
relaxivity, surface 123, 124
relaxometry 4
resolution, spatial 85, 90
resonance condition 18
resonance frequency 3
resonance phenomenon 29
rotating rf field 20
rotation 16, 17, 20, 39, 41, 46, 50, 62
rotation angle 25, 41
rotation matrix 41
rotational ellipsoid 38
rubber 90, 135
rubidium vapor 154
SABRE 155, 157
SABRE polarization transfer catalysts 158
sample rotation 50
saturation 28
saturation, partial 93
saturation factor 149
saturation recovery 28
saturation recovery sequence 112
scattering function 82
Schlumberger 120
Schlumberger tool 121
Schrödinger equation 45, 46
Schulz-Flory distribution 141
second Legendre polynomial 37, 38, 40, 42, 47, 48, 135
self-diffusion 113

self-diffusion coefficient
    114
semi-crystalline polymers
    33, 138, 141
sensitivity 66, 71, 145
sensitivity of NMR 144
SEOP 154
separation NMR 54
separation NMR, 2D 55
signal amplification by
    reversible exchange 155
signal amplitude 127
signal-to-noise ratio 144
signal, complex 32
sinc 25, 87
single-point acquisition 102
single-point imaging 85,
    97
skin 134
slice 87, 88, 89, 109
slice definition 87
slice-selective 2D imaging
    91
slice-selective pulse 89
slow-motion approximation
    104
slow-motion limit 62
small flip-angle pulse 92
solid echo 49, 112
solid effect 148, 150, 151
solid fat content 133
solid-state echoes 49
solid-state NMR spectra,
    high-resolution 51
solid-state NMR spectro-
    scopy, high-resolution ^{13}C
    5
solvent cleaning 142
solvent exposure 141
space encoding 101
sparse sampling 86
spectral editing 131
spectroscopy 4, 83
spectroscopy, 2D NMR 54
spectroscopy, 2D J-resolved
    55
spectroscopy, 2D NOESY
    130
spectrum 23, 24, 32
spectrum, 2D 54, 59
spectrum, 2D J-resolved 56

spherical harmonic functions
    37
SPI 85
spin 13
spin coupling, asymmetric
    40
spin density 31, 76, 83,
    93
spin-exchange optical
    pumping 154
spin order 156, 157, 158
Spinlab 152
spinning sidebands 50
spinning top 16
SQUID 145
stable radicals 150
steady-state free precession
    96
stiffness, chain 135
stochastic resonance 29
stray-field NMR 115
stray-field NMR relaxometer
    132
stress-strain curve 140
structure, secondary 66
sub-spectra 68
surface-to-volume ratio 113,
    123
susceptibility, magnetic 93
susceptibility difference,
    magnetic 115, 123
susceptibility differences
    113
susceptibility effects 93
suspensions 133
T_1 weight 126
Taylor expansion 78
Taylor series 77
temperature map 94
tertiary structure 66
thermal mixing 148, 151
thermodynamic equilibrium
    144
thermodynamic equilibrium
    density matrix 46
thermodynamic equilibrium
    magnetization 17, 27, 28
thermodynamic equilibrium
    polarization 58
time domain 112
time-domain NMR 112

time-evolution operator 46
tire 90
tire testing 137
tissue, biologica 134
TOCSY 60
tomography 4
tortuosity 125
transfer of longitudinal
    magnetization 67
transfer of spin order,
    physical 157
transfer path 158
transition frequencies 45
transmitter 18, 21
transport phenomena 100
transverse magnetization
    16, 22, 26, 28, 43, 46
transverse relaxation 64,
    90, 93
transverse relaxation decay
    138
transverse relaxation rate
    113
transverse relaxation time
    16, 22
tri-exponential function 138
two-dimensional NMR 5
unilateral NMR 4, 5
unpaired electrons 146
velocity 77, 80, 107
velocity, average 106, 108
velocity distribution 83,
    103, 104, 117
velocity distribution, 2D 104
velocity encoding 85
velocity image 103
velocity imaging 101
velocity profile 103
velocity-exchange NMR 108
velocity-exchange
    spectroscopy 106
velocity-vector fields 100
VEXSY 106, 108
viscosity 128
viscosity estimates, oil 122
volume-selective
    spectroscopy 88
von Neumann equation of
    motion 46
vortex motion 102
voxel 74

vulcanization 90, 135
WAHUHA sequence 52
wall relaxation 113, 116,
    121, 127
water, bound 125
water, capillary-bound 122
water, clay-bound 122
water drop 102, 107
water pipe 107
water salinity 127
water saturation, irreducible
    122
wave functions 45
wave number 82
wax 141
weight function 132
weight parameter 139
well logging 119, 120, 123,
    126
well logging NMR 5, 113
wideline exchange NMR
    spectra 62
wideline NMR spectroscopy
    48
wideline NMR spectrum 47,
    49
wire-line mode 120
X-ray diffraction 140

Printed in the United States
By Bookmasters